SIMPLICIUS

On Aristotle's

"On the Heavens 1.5-9"

SIMPLICIUS

On Aristotle's "On the Heavens 1.5-9"

Translated by
R. J. Hankinson

Cornell University Press

Ithaca, New York

First published 2004 by Cornell University Press.

ISBN 0-8014-4212-5

Acknowledgments

The present translations have been made possible by generous and imaginative funding from the following sources: the National Endowment for the Humanities, Division of Research Programs, an independent federal agency of the USA; the Leverhulme Trust; the British Academy; the Jowett Copyright Trustees; the Royal Society (UK); Centro Internazionale A. Beltrame di Storia dello Spazio e del Tempo (Padua); Mario Mignucci; Liverpool University; the Leventis Foundation; the Arts and Humanities Research Board of the British Academy; the Esmée Fairbairn Charitable Trust; the Henry Brown Trust; Mr and Mrs N. Egon; the Netherlands Organisation for Scientific Research (NWO/GW). The editor wishes to thank Catherine Dalimier, Stuart Leggatt, Andrea Falcon, Istvan Bodnar and Henry Mendell for their comments, and Ian Crystal and Han Baltussen for preparing the volume for press.

Printed in Great Britain

Librarians: Library of Congress Cataloging-in-Publication Data are available.

Contents

Conventions

[…] Square brackets enclose words or phrases that have been added to the translation or the lemmata for purposes of clarity, as well as those portions of the lemmata which are not quoted by Simplicius.

<…> Angle brackets enclose conjectures relating to the Greek text, i.e. additions to the transmitted text deriving from parallel sources and editorial conjecture, and transposition of words or phrases. Accompanying notes provide further details.

(…) Round brackets, besides being used for ordinary parentheses, contain transliterated Greek words.

Preface

Richard Sorabji

Aristotle argues in *On the Heavens* 1.5-7 that there can be no infinitely large body, and in 1.8-9, following Plato *Timaeus* 31B-33B, that there cannot be more than one physical world. As a corollary, in 1.9, Aristotle infers that there is no place, vacuum, or time beyond the outermost stars. Simplicius the Neoplatonist wrote his commentary on the text in the sixth century AD.

As one argument in favour of a single world, Aristotle (1.8, 276a18-b21) defends his system in which there are no more than four natural destinations for the four elements, viz. earth, water, air and fire. Earth is naturally located at the centre of his concentric universe, and this counts as down. Fire is naturally at the periphery of the part beneath the heavens, and this counts as up. Water and earth have intermediate natural destinations. If these elements cannot have more than one natural destination each, there cannot be more than one world formed out of them.

Soon, however, Aristotle's own school, under the third head, Strato, was to challenge this system, so Simplicius reports: 267,29; 269,4. Strato thought that, instead of having natural places, the four elements all move towards the centre, but the lighter ones get squeezed outwards. Others, like the atomists, allowed bodies to move throughout an infinite universe, but Aristotle goes on to argue, in 1.8, 277a 27-33, that the elements accelerate as they approach their natural places, and acceleration must be across finite distances, if speed is not, absurdly, to become infinite.

In this context, Simplicius tells us of several ancient theories of acceleration. Hipparchus, the astronomer of the second century BC, is reported at 264,25-265,9 as having explained the acceleration of falling bodies as due to the decline of an initial force opposing fall, impressed in the body by what threw it up (the earliest example of what came to be called 'impetus'), or by what detained it aloft. Simplicius 264,22-5; 266,29-267,6, ascribes to Aristotle, following Alexander, the different view that the acceleration of falling bodies is due to the form of earth getting perfected as it nears its destination. But he objects that the same body would register different weights when weighed aloft and when near its destination, and he wonders whether the acceleration occurs at all. A third theory, reported at 266,3-17, explains the acceleration of falling bodies as due to the greater resistance of the larger body of intervening air to rocks falling from higher up.

Aristotle offers another argument for there being only one world in *Metaphysics* 12.8, 1074a31-8, that there is only one divine mover of the heavens. Alexander, at 270,9-12, is reported as being puzzled as to why one mover should not move more than one world, especially as he moves it like someone inspiring love, *Metaphysics* 12.7, 1072b3. Simplicius replies for one thing, 271,13-21, that his teacher Ammonius had shown that in Aristotle's view, the heavens are beginninglessly produced, as well as beginninglessly moved, by God, and Alexander has ignored this. This interpretation artificially aligns Aristotle with Plato, who, on the Neoplatonist interpretation, also makes his God create beginninglessly. According to Simplicius, Aristotle intended his argument in *Physics* 8.10, to apply to producing existence as well as to producing motion, when these have neither beginning nor end. For both equally, an infinite power is needed, and the infinite power has to be an immaterial being, since infinite power cannot be housed in what is finite, and all bodies are finite in extent. Thus Aristotle's proof of an immaterial mover of the heavens is represented as equally a proof of an immaterial source of existence.

Alexander further overlooks, says Simplicius at 270,14-27, that Aristotle postulates more than one divine mover, because the movements of the lower spheres which carry the planets are inspired by separate unmoved movers, and he cites Aristotle *Metaphysics* 12.8, 1073a23-34.

But here we run into a question about Alexander's simplification of Aristotle's astronomical system. Alexander seems to have reduced the number of spheres that carry celestial bodies from 55 in Aristotle to 7, and when he refers to the 7 unmoved movers of these spheres, he might be referring to their souls rather than to divine movers, other than the supreme one, inspiring them with love.

The case for the 7 spheres of Alexander all being inspired just by the one divine mover has been put by István Bodnár, 'Alexander of Aphrodisias on celestial motions', *Phronesis* 42, 1997, 190-205, and although the evidence is not conclusive, it is suggestive. Bodnar's first piece of evidence is our passage, in which Alexander wonders why all spheres should not love the same unmoved mover, ap. Simplicium *in Cael.* 270,9-12. Secondly, Alexander is reported as arguing that souls which are spread through rotating bodies qualify as not being moved even incidentally, since the rotating bodies themselves do not as a whole change their place (ap. Simplicium *in Phys.* 1260,22-35). Thirdly, Alexander is willing to admit after all that the souls in the inner spheres are moved incidentally, insofar as these spheres undergo a second movement which is produced not by (*hupo*) their souls, but, presumably, by the outermost sphere. But in this context, he is presented as thinking that the movement produced by their souls is produced by (*hupo*) unmoved entities, which, being within them (*en autois*), sound like souls. Moreover, he is presented as holding that the mover of the outermost sphere is a separated substance, to be contrasted with all other souls which are inseparable from bodies, ap. Simplicium *in Phys.* 1261,30-1262,4. Fourthly, in *On the Cosmos* translated from Arabic by Genequand, Leiden 2001, at 86, Alexander admits a distinct mover and desiring element in each sphere, but at the same time suggests, for a reason he may not eventually accept, that it is 'probably' wrong to admit a plurality of movers, presumably movers of another kind, and so perhaps movers acting as objects of love.

As regards Aristotle's next question, the finitude of place and the absence of place beyond the outermost stars, Simplicius ascribes to the Stoics at 284,28-285,2, the argument elsewhere attributed to Plato's Pythagorean friend Archytas in the fourth century BC. There cannot be an outermost edge to the universe, for imagine trying to extend your hand beyond it. If you succeed, there is more space beyond the 'edge', whereas if you fail, there is more matter beyond the 'edge', which is blocking you. Alexander is reported at 285,21-7 as replying that what stops you is the sheer non-existence of anything, whether body or place, beyond the outermost stars. For you cannot extend your hand into the non-existent. Aristotle's own account of a thing's place makes it in effect the inner surface of its immediate physical surroundings, and this would give the result that Alexander alleges, that there is not even place, given Aristotle's astronomy, because there are no physical *surroundings*, beyond the stars.

Alexander is reported as objecting to the thought experiment with the hand by saying that imagination is not a good guide to what is possible, and that one should not rely on impossible thought experiments. A person could not enter the region of celestial matter. But Simplicius reminds him that Plato and Aristotle certainly made use of such impossible thought experiments, 285,5-21; 286,23-7. When Aristotle imagines stripping away all form in order to reveal the nature of first matter, the impossibility of form existing without matter is of an even stronger kind than the impossibility of a human entering the heavens. Impossible thought experiments were used not merely for reducing to absurdity, but for investigating the nature of things, and were referred to as *hypotheses* at least by the time of Aristotle's pupil Eudemus, reported by Boethius *On Hypothetical Syllogisms* 1.2.5-6, PL 64, 833D. See the discussion in Christopher Martin, 'Non-reductive arguments from impossible hypotheses in Boethius and Philoponus', *Oxford Studies in Ancient Philosophy* 17, 1999, 279-302.

Aristotle himself argues against there being any place beyond the furthest star, by insisting at 1.9, 279a11-18, that place and vacuum are defined as what *can* receive matter, but since matter cannot get beyond the outermost stars to be received, there is not anything there that *can* receive them. The most devastating reply to this argument is not given by Simplicius, but is provided by the Stoic Cleomedes, *de motu circulari*, Todd 10,6-15, who says he gets most of his arguments from Posidonius. According to Cleomedes, you might as well say that there cannot be a water vessel in a desert that cannot receive water, since a water vessel by definition *can* receive water.

When Aristotle says that neither place nor time belong to the things beyond the heavens, *On the Heavens* 1.9, 279a17-b3, Alexander takes him to be referring to the outermost sphere which carries the non-planetary bodies, according to Simplicius 287,19ff. But Simplicius 290,1ff. prefers the view that the reference is to the divine intellect, or intellects, which, on Aristotle's theory, inspire movement in the heavens. In that case, Aristotle might be saying that they are altogether divorced from time. Elsewhere he talks ot things being timeless only because, as everlasting, they are not *measured* by time.

*

A new introduction to the commentators will appear in Richard Sorabji (ed), *The Philosophy of the Commentators, 200-600 AD: A Sourcebook*, 3 vols, forthcoming 2004 (London: Duckworth).

Introduction[1]

(i) Simplicius: life, times and work

Very little can be known for certain about Simplicius' life. He was born in Cilicia in Asia Minor, probably towards the end of the fifth century AD. He attended lectures on philosophy given by the Neoplatonist Ammonius in Alexandria, as did his great philosophical nemesis John Philoponus, the Christian Neoplatonist and antagonist of Aristotle's cosmology, polemic against whom occupies much of Simplicius' commentary on Chapters 1-4 of Book 1 of *de Caelo*.[2]

Philoponus probably lived from *c.* 490 – *c.* 570, which would place him with Ammonius sometime around 510 (Ammonius' teaching career spanned some fifty years, straddling the turn of the sixth century, and so is as such of little help in determining the dates of his pupils).

It appears from some remarks of Simplicius that he never actually encountered Philoponus in person (*in Cael.* 26,18-19: he says that his disdain for Philoponus has nothing personal in it), a fact which, in view of the intemperateness and vehemence of his animosity towards him, may well be accounted fortunate. On the assumption that Philoponus would have been sitting at Ammonius' feet in the years around 510, Simplicius must have attended Ammonius' lectures either shortly before or shortly afterwards. Philoponus produced at least some of his first major work, *On Aristotle's 'Physics'*, in 517, while *in de Caelo* cannot have been written before 529 (the year in which Philoponus' polemic *Against Proclus on the Eternity of the World* was published; shortly afterwards came *Against Aristotle on the Eternity of the World*,[3] against which Simplicius responds at such outraged length in *in Cael.* 1.2-4).[4] On the (evidently fragile) basis of the fact that Philoponus appears to have published earlier, we might suppose that Simplicius studied with Ammonius in about 515.

He also studied with Damascius, the last head of the school of Platonic philosophy in Athens, presumably after his Alexandrian sojourn, who was to go into exile with him when the school, often, if probably inaccurately, referred to as the Academy,[5] was shut down, along with the other pagan schools of philosophy, by the Emperor Justinian in 529.

This is the traditional date of the end of Greek philosophy. But

Simplicius was a Greek philosopher in every sense – and he had probably hardly begun to write at the time of Justinian's edict. I say 'Greek in every sense', because he lived and breathed the atmosphere of Greek classical thought, as it was filtered through the screen of pagan Neoplatonism. His world was the world of Plato and Aristotle, and of the great tradition of commentary and exegesis and development that stretched back to Alexander of Aphrodisias at the beginning of the third century AD, and beyond to Aspasius and earlier Peripatetics such as Xenarchus and Nicholas of Damascus, both of whom are quoted in *in de Caelo*.

And although the Roman Empire, both Western and Eastern, had been Christian (with the occasional hiccough) since the reign of Constantine at the beginning of the fourth century, pagan philosophy had continued to flourish, even in strongholds of the new religion such as Alexandria, in relative peace and quiet, at least from the time of Plutarch of Athens onwards (*c.* 400). Simplicius' outlook was the product of that tradition; and he considered Christianity to be a vulgar, intellectually disreputable, ephemeral upstart. In an eloquent, if embittered passage, he likens the current success of Christianity (and with it the writings of its avatar Philoponus) with the luxuriant but fleeting blooms in the garden of Adonis (*de Caelo* 1.2, 25,34-6).

Not that he was unreligious: on the contrary, his work is suffused with a genuine religious reverence for the author of the universe, whom he conceives, in Platonist rather than Neoplatonist fashion, as the cosmic Craftsman. The last words of *in de Caelo* take the form of a dedicatory prayer:[6]

> I offer these things as a hymn to you, master and artificer both of the entire universe and of the simple bodies within it, and to your creatures, desiring to contemplate the greatness of your works and to make them apparent to those who are worthy, so that, without thinking that there is anything shoddy or human about you, we may adore you in accordance with the transcendence which you possess in relation to everything which you have created. (*in Cael.* 4,6, 731,25-9)

This was genuine spirituality, in sharp contrast to what he saw as the meretricious counterfeit of Christianity.

But that meretricious counterfeit had had more than two centuries to establish itself at the heart of Roman and Byzantine culture by the time Justinian expelled the pagans. By Simplicius' day, Byzantium was the Roman Empire, the Western Empire having finally guttered out in 475 when the Goths pushed the last, pathetic Emperor, the boy Romulus Augustulus, aside, and established a Christian Gothic kingdom in Italy first under Odoacer and then Theodoric, who was to commit his own crime against philosophy by having the Christian Senator Boethius imprisoned, tortured and garrotted in 524.

It is not clear whether Justinian actually exiled the non-Christian philosophers after the closure of their school. He was certainly by no means uniformly hostile to pagans. In the same year 529, on April the 8th to be exact, the pagan Tribonian published the *Codex*, the new systematization of Roman Law, and shortly afterwards followed it up with the *Digest* (or *Pandects*) of 530 and finally the *Institutes* of 533 – all work he had carried out at Justinian's behest, even though he was, with the possible exception of the rapacious if nominally Christian Praetorian Prefect John of Cappadocia, the most hated man in the Empire, because of his notorious judicial venality in his office of Quaestor, the equivalent of the Attorney General.

At all events, the philosophers left Athens: 'Damascius of Syria, Simplicius of Cilicia, Eulamius of Phrygia, Priscian of Lydia, Hermias and Diogenes from Phoenicia, and Isidore of Gaza, the finest flower … of our time' (Agathias *Histories* 2.30.3) decamped for the safety of the Persian Empire. The Persian King Chosroes I, who ascended the throne in 531, was an enemy of Byzantium, and was to remain so for the rest of his long reign. But at the signing in 532 of a peace-agreement with Justinian, whose attentions were turning to the reconquest of Roman North Africa and eventually the whole of Italy from the Goths under his brilliant general Belisarius, Chosroes insisted on adding a clause providing for the philosophers' safe return to Constantinople, should they wish to avail themselves of it.

Until recently, it has usually been assumed that they did indeed return, probably to Athens, although never again to teach Platonic philosophy with the imprimatur of an active institution. But in the past few years, Michel Tardieu has argued that Simplicius at least did not return, but settled in the town of Harrân, near to, but on the Persian side of, the Byzantine frontier.[7] Ingenious and well-argued though it is, Tardieu's thesis has not commanded universal assent;[8] and his contention that there was a flourishing school associated with Simplicius in Harrân has been exploded.[9]

None the less, it is far from clear that Simplicius ever did return – at all events, there is no positive evidence in favour of the theory – and school or no school, he may well have remained in Harrân, which Chosroes deliberately built up as a redoubt against the influence of Christianity: Procopius, in the official and less entertaining part of his history, record that Chosroes exempted the citizens of Harrân from taxes as a reward for their staunch commitment to paganism (*Bella* 2.13.7); Simplicius would surely have felt at home there.

But even if he did return to the Byzantine Empire, it would have been to a life of intellectual exile; he would never again enjoy the convivial company of like-minded thinkers in an institutionally sanctioned atmosphere. From now on, in the Byzantine world, officially at least philosophy was theology; and the theology was Christian. Simplicius would have found himself an intellectual stranger in an intellectually alien land. He probably died some time in the 550s, the decade that saw the death of Justinian, and the restriction of Byzantine power in Italy to the Exarchate of Ravenna.

(ii) Simplicius: *in de Caelo*

As noted above, Philoponus' *Against Proclus on the Eternity of the World* was published in 529 (its date is established by an astronomical reference: *Against Proclus* 579,14-17);[10] *Against Aristotle* was written later,[11] although probably not much later. *In de Caelo* is almost certainly (on the basis of internal cross-references) the earliest of Simplicius' surviving works, antedating *in Physica*,[12] which was certainly written after 532, and likely after 538, since it speaks of Damascius as though he were already dead, while he is known to have been alive at the earlier date and probably was at the later.[13] This means that *in de Caelo* was probably composed some time in the 530s – but more than that it is impossible to determine.[14]

The most obvious feature of *in de Caelo* is its length. Aristotle's *de Caelo* is a work in four books, of uneven length (the first two, dealing with the structure of the cosmos and the nature of the heavenly bodies, take up about two-thirds of the text). It runs to 45 pages in the Bekker edition which is standardly used for reference. To comment on that, Simplicius uses 731 pages of the large-format Prussian Academy edition of Heiberg. Allowing for the difference in page sizes, this yields a commentary to text ratio of about 10:1. This figure is somewhat inflated by the fact that Simplicius devotes 200 pages to Chapters 1-4 of Book 1 of *de Caelo* alone (a slab of text which takes up less than three full Bekker pages), a degree of prolixity accounted for in part by the density and importance of the material, but primarily by the fact that Simplicius allows himself lengthy digressions to attack Philoponus' refutation of Aristotle, and also, *inter alia*, to offer a 15-page disquisition on Neoplatonist metaphysics.

Simplicius also quotes at length from earlier writers, in particular the commentary of Alexander of Aphrodisias on *de Caelo*, a work which he respects, although he frequently takes issue with it (that he does so respectfully is witnessed by the frequency with which Simplicius will preface his objections with a cautious 'perhaps'). In the commentary on *de Caelo* 1.2, he preserves extensive quotations from the lost *Against the Fifth Substance of Aristotle* of Xenarchus, an unorthodox Peripatetic of the first century BC.[15]

All of this, naturally, adds to the length of the work; but even so, a modern reader might be forgiven for thinking Simplicius' commentary excessively prolix even when it concentrates on the exegesis of Aristotle's own words. A modern reader will probably also find, at least some of the time, the issues chosen for concentration strange as well. Much of this strangeness is dictated by the very different intellectual world in which Simplicius was writing.

He was, as has been said, a Neoplatonist. That is to say, he accepted as being the correct account of the metaphysical structure of things the complex hierarchical ontology pioneered by Plotinus (AD 204-270), and developed by subsequent followers, notably Porphyry, Iamblichus, Proclus and Damascius. Plotinus, while accepting that there was much of value in Aristotle, explicitly rejected his scheme of categories in three related treatises (*Enn.* 6.1-3). His pupil and biographer Porphyry, however, defended the Aristotelian scheme in two commentaries on the *Categories*, one of which survives.

The main issue was the extent to which Aristotle's views could be harmonized with Platonic transcendentalism. At first sight, given Aristotle's well-known attacks on the Theory of Forms, such a project might seem hopeless, crazy even – and indeed it has been castigated as such.[16] And the proposition that the two great Masters of Classical Greek philosophy were fundamentally in agreement did not command universal respect, even among Neoplatonists. Porphyry had merely said, *contra* Plotinus, that the Aristotelian categories could be applied to the transcendental world of the Forms; his most important pupil Iamblichus (fl. *c.* 300), on the other hand, held that Aristotle did not disagree with Plato about the Forms at all (Iamblichus, in Elias *in Cat.* 123,1-3).

Syrianus and his pupil Proclus (412-85), whose *Elements of Theology* provides then most elegant and orderly synthesis of Neoplatonist metaphysics, while accepting that there was a deep concinnity between the two, could not bring themselves to go quite this far – the Forms were an unavoidable stumbling-block to the thesis of universal harmony, but equally problematic were the apparent divergences in the Masters' cosmologies.

The Neoplatonist curriculum, which by this time had become fairly strictly institutionalized, began with Porphyry's *Introduction* to the *Categories* (one of several such Neoplatonic propaideutic *Introductions*, some of which, especially Iamblichus', bore a close relation to Aristotle's lost *Protrepticus*) and the *Categories* themselves, then proceeded through the rest of Aristotle (the so-called Lesser Mysteries) to Plato (the Greater Mysteries), the capstone of which was the *Timaeus* and the *Parmenides* (the latter being read as an essay on the Neoplatonic One), followed (it was hoped) by a final ascent to mystical union with the One of the sort which Plotinus claimed to have achieved on a few occasions.

Such a curriculum will strike most contemporary students of (western) philosophy as a trifle odd for a number of reasons, not least in the nature of that final goal, although it will no doubt seem familar to adepts of eastern wisdoms. But equally peculiar to modern eyes is the notion, implicit in the curricular structure, that Plato's most important dialogues are *Timaeus* and *Parmenides*; the former is not the strange, and in parts somewhat embarrassing sketch of a cosmology that so many modern critics have seen it as (although in recent years there have been refreshing signs that the dialogue is being taken more seriously); the latter not even Gilbert Ryle's early essay in the Theory of Types, much less the 'primer of elementary fallacies' that some have discerned.

The centrality to Neoplatonic Platonism of the *Timaeus* is of particular importance for our purposes. As noted above, one of the sticking-points for the universal harmony thesis was cosmology. Plato seems to describe in the *Timaeus*, albeit only in the form of a 'likely story', a generation of the cosmos by a designer-god (the Demiurge) on the basis of a perfect exemplar, utilizing matter that was in some sense already there, and deliberately creating a structure in accordance with the best possible design.

None of this seems remotely compatible with the views expressed by Aristotle, pre-eminently in *de Caelo*; for him the universe is uncreated and eternal (*Cael.* 1.10-12); God has no causal role (apparently) in its continued operations, but merely functions as a focus of admiration and emulation (*Cael.* 2.12); and hence, while nature does nothing in vain, it does not do so as a result of any conscious decision to pursue the best possible outcome. Teleology is immanent in the nature of things, not imposed upon them from the outside.

All of this, added to Aristotle's championing of the existence of a fifth element for the heavenly bodies (*Cael.* 1.2-4), for which there seems no Platonic antecedent (and which drew the fire of both Xenarchus and Philoponus, much to Simplicius' righteous indignation), and to the difficulties involving the Forms, make the project of reconciliation look particularly hopeless in these contexts. Neoplatonists were inclined to interpret the Forms as ideas in the divine Mind; and it seems impossible to reconcile such a view with what Aristotle says about form and against Forms.

But Proclus' pupil Ammonius found a way to attribute just such a doctrine to Aristotle (Asclepius *in Metaph.* 69,17-21; 71,28, reporting Ammonius); and gradually the universal harmony thesis came to be the orthodoxy (although it never commanded universal assent). Ammonius also wrote a treatise demonstrating that Aristotle's God was indeed an efficient cause of the cosmic ordering, not indeed in the sense that he created it in time (not even the Neoplatonists could get Aristotle to say that), but in the sense that he is constantly efficiently involved in its conservation (for the Neoplatonists, indeed for most of the ancients, conservation was just as much a manifestation of efficiency as production – as of course it was to be for Descartes). This treatise does not survive; but his pupil Simplicius summarizes its arguments at *in Phys.* 1361,11-1363,12, and was clearly deeply influenced by it.

For Simplicius was, as far as the harmony thesis went, the purest of the pure. At the beginning of his own commentary on the *Categories* (*in Cat.* 7,23-32), he states that it is the duty of the commentator to exhibit that harmony to the greatest possible extent – indeed, his abandonment of this duty is one of the things that makes Philoponus anathema to him.[17] Even there, in stating his manifesto, Simplicius makes it clear that there are limits to the feasibility of such an enterprise – the commentator should only seek to do so 'in most cases': inevitably there will be passages that cannot be harmonized.

But there can be no doubt of Simplicius' overall commitment to the project: and it proves the source of many of his disagreements with Alexander (including the question of whether God – the Prime Mover – is an efficient cause: *in Cael.* 1.8, 269,31-271,27).

It also drives his disputes with many others quoted and referred to in the text; but more important in this regard, and what might appear to be at odds with the nature of the reconciliation project outlined so far, is Simplicius' commitment to a friendly and supportive exegesis of the details of Aristotle's physics, in particular to a defence of the postulation of the ether, the fifth element, in *in de Caelo* 1.2-4,[18] in the course of which he quotes, refers to, and takes issue with a number of his illustrious predecessors, pre-eminently Xenarchus and Philoponus.

In the end, what is striking about Simplicius' commentary, for all its verbosity and prolixity, and for all that it spends much time embroiled in *minutiae* in a manner that will tax the patience of most modern readers, and for all the implausibility – from a modern standpoint[19] – of his guiding exegetical principle, is how sensitive a reader of Aristotle he turns out to be. When he disagrees with Alexander on a point of interpretation, he is, in my view, more often right than not (the interpretation of the nature of the Prime Mover is, for me at least, an exception). And that is reason enough for the serious student of Aristotle to take him seriously.

But even if Simplicius' writings were adjudged otherwise worthless, they would still be immensely valuable for what they contain of other writers, for *inter alia* they are a thesaurus of fragments and testimony of ancient writings otherwise lost. To cite only the best-known example, virtually all that we possess of Parmenides' 'Way of Truth' is preserved in Simplicius' commentary *On the Physics*, along with sizeable portions of our extant remains of Empedocles, and virtually all of Melissus. *In de Caelo* preserves invaluable fragments and testimony of Xenarchus and Philoponus, but also of Iamblichus, Nicholas of Damascus, and Ptolemy, as well as from the lost works of Aristotle himself. The part of the text translated in this volume contains important information about the great astronomer and scientist Hipparchus[20] (*c.* 180 – *c.* 120 BC), in the course of a penetrating discussion of various rival ancient theories of weight.[21]

But there is more to Simplicius than just that. Whatever our attitude to the (almost) universal harmony thesis, exposure to it is invaluable as a means of jolting us from the unreflective comfort of our modern predilections. And this modernity should be stressed; the Neoplatonists were not the first to adopt a syncretizing approach to the works of Plato and Aristotle – that had already been a commonplace of the earlier Middle Platonic tradition, of which the *Handbook* of Alcinous is a typical example (see Dillon, 1993) – nor were they the last. And even if, at the end of the day, we decide, with most modern interpreters, that the syncretizing project of reconciliation is a non-starter, our understanding of the immensely complex relations that hold between the philosophies of Plato and Aristotle will have been deepened by our having had the chance to confront our judgements and prejudices by viewing them through an utterly different glass, albeit sometimes darkly. And – who knows? – we may even come to see it as something less than utterly crazy.

(iii) Text and translation

This volume is the second of three, which together translate the commentary sections of Simplicius' commentary on the first book of *de Caelo* (the first and third translate his Chapters 1-4 and 10-12 respectively). In the part of the book translated here, Chapters 5-7 deal with the question of whether the universe could be infinite in extent, that such a supposition is incoherent is argued on geometrical grounds in Chapter 5, and on physical grounds in Chapter 6. Chapter 7 concludes with a refutation of the atomist hypothesis of an infinity of atoms in an infinite void. Chapters 8 and 9 argue for the uniqueness of the finite universe.

This division was necessitated by the length of Simplicius' text – but it is not entirely an arbitrary one. At the end of his *Prologue* (6,7-27), Simplicius himself offers a résumé of the contents of Book 1 of *de Caelo*, which suggests the naturalness of these breaks. Thus the resulting three volumes deal with the basic subject-matter of the inquiry into nature, and the argument for the existence of a separate element for the heavenly bodies (*Cael.* 1.1-4), the finitude and uniqueness of the cosmos (*Cael.* 1.5-9), and its ungenerability, indestructibilty, and eternity (*Cael.* 1.10-12).

After the opening *Prooemium* (1,1-6,27: translated in Hankinson, 2002), Simplicius treats Aristotle's text line by line, lemma by lemma, chapter by chapter. The MSS of Simplicius reproduce only the opening and closing words of each lemma, the supposition presumably being that anyone using the commentary would also have a text of Aristotle to hand. In common with some of the other translators in this series, I have chosen to include the whole of the texts that Simplicius comments upon: the parts of each lemma that do not appear in the MSS are enclosed within square brackets.

This volume thus contains, *inter alia*, a translation of part of the first book of *de Caelo*. The basis for this translation is the version prepared by Mohan Matthen and myself for our forthcoming collaborative contribution to the Clarendon Aristotle Series, but it is not identical with it. I have standardized the technical terminology to conform to the translations I prefer in the case of Simplicius; and additionally there are several places where the text of Aristotle that Simplicius had before him clearly differs in some way (usually fairly insignificantly) from the preferred modern versions: D.J. Allan in the OCT, Guthrie in the Loeb Classical Library (*Aristotle*: in twenty-three volumes; VI: *On the Heavens*, London, 1939), or Moraux in the Budé (*Aristote: du Ciel*, Paris, 1965). I have indicated these divergences in notes as they occur.

The Greek text used is that of I.L. Heiberg in the *Commentaria in Aristotelem Graeca* (*CAG*: under the general editorship of H. Diels), vol. VII, Berlin, 1894. In a few places where I have departed from that text, I have noted the fact ad loc., and a full list of preferred readings appears on pp. 13-14.

I am much in the debt of several anonymous 'vetters', as a result of whose veterinary skills my translation is much healthier than it would otherwise have been, in terms of both accuracy and felicity. On the occasions where I have rejected their advice, I have not done so lightly or without careful consideration. The Project's various assistants (Han Balthussen, Sylvia Berryman, Elena Vambouli, Ian Crystal) have, at various times, rendered assistance with exemplary skill and tact. Finally, I would also like to thank Richard Sorabji, for inviting me to contribute to his series, for his unfailing enthusiasm and encouragement, for several particular suggestions for improvement which I have gratefully adopted, and above all for his patience in awaiting the results of a project that took me much longer than I had originally expected it to.

Notes

1. This volume is the second of three, translating Simplicius' commentary on Book 1 of Aristotle's *de Caelo*; the first volume (Hankinson, 2002) contained an extensive general introduction, dealing with Simplicius' life, work and times, and with the impact of *de Caelo* on the subsequent history of thought. Rather than simply reproduce that here, this introduction provides an abbreviated account of Simplicius' life and works, before turning to specific features of the chapters here discussed.

2. See Hankinson, 2002, for further details. The commentary, *in Aristotelis de Caelo Commentaria* (hereafter *in Cael.*) is edited by I.L. Heiberg, as vol. VII of *Commentaria in Aristotelem Graeca* (Berlin, 1894).

3. The surviving fragments of which, largely culled from *in de Caelo*, are ably edited by C. Wildberg, *Philoponus: Against Aristotle on the Eternity of the World* (London, 1987).

4. See Hankinson, 2002, Introduction, for further details of the Simplicius-Philoponus feud.

5. It has now been conclusively established that, contrary to long-lived philosophical legend, there was no continuously-operating Platonic Academy in

Athens from the fourth century BC to AD 529; see J. Glucker, *Antiochus and the Late Academy* (*Hypomnemata* 56) (Göttingen, 1978).

6. See on this, and on Simplicius' attitude to Christianity, P. Hoffman, 'Simplicius' polemics', in R. Sorabji (ed.) *Philoponus and the Rejection of Aristotelian Science* (London, 1987).

7. See M. Tardieu, 'Sâbiens coraniques et "Sâbiens" de Harrân', *Journal Asiatique* 127 (1986); *id.*, 'Les calendriers en usage à Harrân d'après les sources arabes et le commentaire de Simplicius à la *Physique* d'Aristote', in I. Hadot (ed.) *Simplicius – sa vie, son oeuvre, sa survie* (Berlin 1987), and I. Hadot, Introduction, *Simplicius. Commentaire sur le Manuel d'Epictète*, Leiden 1996.

8. The doubters include P. Foulkes, 'Where was Simplicius?', *JHS* 112 (1992), 143; S. van Riet, 'A propos de la biographie de Simplicius', *Rev. Phil. de Louv.* 89 (1991), 506-14; H.J. Blumenthal '529 and its sequel: what happened to the Academy?' *Byzantium* 48 (1978), 369-85, and reprinted in his *Soul and the Intellect* (Aldershot, 1993). I deal with the issue at somewhat greater length in the Introduction to Hankinson, 2002.

9. By J. Lameer, 'From Alexandria to Baghdad: reflections on the genesis of a problematical tradition', in G. Endress and R. Kruk (eds) *The Ancient Traditions in Christian and Islamic Hellenism* (Leiden, 1997), 181-91 (Arabic evidence), and recently by C. Luna, review of Thiel 1999, *Mnemosyne* 54-4 (2001), 482-504 (Greek evidence).

10. *Against Proclus* is edited by H. Rabe (*De aeternitate mundi contra Proclum*, Teubner, 1899).

11. This is established by cross-references between the two treatises; e.g. at *in Cael.* 1.2, 135,27-8, = fr. 72 Wildberg, Philoponus is quoted as referring back to *Against Proclus* (which Simplicius had not read: 135,29-31) in *Against Aristotle*; and cf. *in Cael.* 1.2, 136,17-18, = fr. 73 Wildberg; while at *Against Proclus* 258,22-6, Philoponus trails the arguments he is going to write in *Against Aristotle.*

12. However *in de Caelo* does appear to contain a reference to the *Physics* commentary (*in Cael.* 1.3, 108,20: see Hankinson, 2002, n. 155 ad loc.).

13. For the evidence for this, deriving from an epitaph for a dead slave attributed to 'the philosopher Damascius' on a funeral stele now in Emesa dated to 538, see P. Hadot, 'The life and works of Simplicius in Greek and Arabic sources', in Sorabji, 1990, 290.

14. One further suggestive piece of evidence derives from a reference to a first-hand observation of the strength of the wind made near the river Aboras in Mesopotamia (*in Cael.* 2.13, 525,13), which passes close to Harrân: hence it is plausible to date the commentary to the period when Simplicius settled there – but as noted above, all of this speculation is controversial, and controverted.

15. See Hankinson, 2002; and Hankinson, forthcoming 2.

16. By Richard Sorabji: see his General Introduction to earlier volumes in this series.

17. *In Cael.* 1.3, 84,11-15; 1.4, 159,2-9; in the latter passage, he derides Philoponus as an opsimath for not seeing that we must try to uncover the deeper agreements behind the surface inconcinnities; this is just one example of Simplicius' rich and varied vocabulary of abuse, which is liberally applied to Philoponus, who is 'the Grammarian', 'the Telchin', and sometimes simply 'that man'.

18. See Hankinson, 2002.

19. This is worth stressing: see further below.

20. See 1.8, n. 433.

21. *in Cael.* 1.8, 263,13-267,6; the section on Hipparchus runs from 264,25-266,29.

Textual Emendations

210,15	Reading ACE with D, against AEC of the other MSS and Heiberg
212,12	Perhaps omitting *epeidê*
217,13	Replacing ';' with '.'
218,6	Reading *anôthen katô*, with D, in place of Heiberg's *katô*
223,25	Reading *diaretos* in place of *diareta*, Db, Heiberg; other readings: *diairet///* A; *diareton* B
224,13	Omitting *ti meizon heautou* and inserting *to* before *heteron*; or perhaps reading *elatton* for *meizon*
225,5	Perhaps reading *hapasan diastasin peperasmenên* for *hapasan peperasmenên*
227,20	Perhaps omitting *ê menei*
230,9-10	Reading *eiê*, in place of *ara* of the MSS
231,4	Supplying *oude hôs anomoiomeres* after *einai to apeiron*
233,9	Punctuating heavily after *tauta adunata*, and lightly after *kata to deuteron anapodeikton*
233,26-7	Perhaps omitting *kai hôs hê D pros tên E, houtô to Z pros to FB*
235,6	After *tou apeirou* perhaps add *B*
236,6	Supplying *mê en topôi* to complete the lemma (*Cael.* 1.7, 275b11)
240,23	Reading *kinountos* after *tinos*
244,23	Punctuating heavily after *kai ta hexês*
245,17	Closing quotation after *anankê apeiron einai* (quotation marks omitted by Heiberg)
247,8	Reading *tauta* for *touto*
247,13	Perhaps reading *tên diathesin kosmou tinos* instead of *tên tou kosmou diathesin*
248,19	Reading *touto* in place of *to*
250,7	Perhaps reading *eis touto kata phusin estin hê phora autêi*, *vel sim.*, in place of *touto kata phusin estin autêi*
253,18-24	Continuing the quotation from Alexander through these lines, closing the quotation after *apo toude tou mesou*
256,8	Reading *hôs* in place of *hoti*
260,6	Reading *autôn* in place of *autou*
266,26	Perhaps reading *peperasmenou* in place of *apeirou* here

269,6	Reading *pheromenên tautên tên doxan* with b and c, in place of *pheromenên tautên* as printed by Heiberg: but see n. 121 ad loc
272,1	Reading *pros tois oikeiois topois*, with A and D, against Heiberg's *pros tous oikeous topous*
273,4	Retaining *ho* of the MSS, deleted by Heiberg
280,32	Perhaps reading *to sôma tês hulês* with E^2, in place of *tês sômatikês sustaseôs* of Ab, read by Heiberg; or perhaps reading *tês sômatikês hulês*
290,15	Omitting *ta* before *ourania*
291,1	Reading *exêrtêtai*, with E^2c against *epeisin* of A, printed by Heiberg; DE offer *epanietai*
291,11-12	Treating *hoti to theion kai ametablêton anankaion einai pan to prôton kai akrotaton* as a quotation

Simplicius
On Aristotle
On the Heavens 1.5-9

Translation

Simplicius' Commentary on Book One of Aristotle's 'On the Heavens'

[CHAPTER 5]

271b1-17 Since we are clear on these points, then, we should 201,11
examine what remains,[1] [first of all whether there is any infinite body, as most of the ancient philosophers thought, or whether this is an impossibility. For whether it is thus or otherwise makes no small difference, but rather all difference in the world towards the investigation of the truth. For this has been, and will be, the origin of pretty well all the contradictions among those who make some claim about nature as a whole, given that a small deviation from the truth at the beginning multiplies itself ten thousand-fold. For instance, suppose someone were to claim that there was a smallest magnitude: this man would upset the greatest truths of mathematics by introducing his smallest [thing]. The cause of this is the fact that the starting-point is greater in potential than it is in magnitude, so that something small at the beginning in the end becomes gigantic. But the infinite has both the capacity of a starting-point, and the greatest quantitative capacity, so it is neither absurd or irrational that the most remarkable difference follows from assuming that there is some infinite body.] Therefore we should speak about this, starting from the beginning.

Alexander says that what is at issue here is to investigate whether the whole world is infinite or finite, since that was what he [sc. Aristotle] proposed at the beginning when he said 'we may inquire later on, then, concerning the nature of the universe, whether it is infinite in respect of magnitude or whether the whole mass is finite'.[2]

Seeking to prove its finitude on the basis of [the fact that] none of the simple [bodies] out of which it is constituted is infinite, since he was unable to show that the fifth body was not infinite without having first spoken of its substance and proved that there is some fifth body, for this reason he returns to the issue after [providing] those arguments,[3] and proves that the five simple bodies are also finite in magnitude. Since he has already clearly shown, on the basis of their motions, that the five simple [bodies] are finite in number, by now showing that they are finite in magnitude [he will show] that the whole world is finite in magnitude.

Alexander, however, believing that the whole treatise is primarily

concerned with the world as a whole,[4] accordingly understands these things too[5] in relation to the study of the universe. But if what I said earlier[6] was true, namely that in this treatise his aim is primarily to treat of the simple bodies, after studying the principles from which the natural bodies are constituted (and primarily the simple ones), he then sets out to investigate infinitude and finitude, primarily in
202,1 relation to the simple elements.[7] And he shows that they are finite as regards both number and magnitude, and then justly supposes in addition that the whole world, being composed of these simple [bodies] will itself be finite.

Properly reducing the argument to more general [terms], he asks 'whether there is any infinite body …, or whether this is an impossibility'. And having assumed by way of necessary division[8] that if there is some body infinite in magnitude it will be either simple or composite, and having shown that none of the simple [bodies] is infinite in magnitude, then, since he already has [the conclusion that] the simple bodies are finite in number, he draws the necessary consequence, as I said, that the composite whole too is finite.

And he first shows that the argument which determines whether or not there is some infinite body is necessary for the natural scientist, from the [fact that] this difference is the cause of practically every disagreement among natural philosophers.

Because of this some (namely those who, like Plato and Aristotle, would not admit the infinite as a principle) said the world was unique and finite, while others, like Anaximenes[9] who said that infinite air was the principle, said there was a single infinite. Others still said that there were an infinite number of worlds, like Anaximander,[10] who posited something infinite in magnitude as a principle and who seems to have generated an infinite number of worlds out of it, and Leucippus and Democritus,[11] who said that an infinite number of worlds were constituted in an infinite void out of an infinity of atoms. And one may say that to posit or not posit an infinite is the origin of all of these oppositions, because the worlds encompass all other oppositions.[12]

Moreover, on account of this dispute some did away with generation, saying like Anaxagoras[13] that everything was precipitated by [a process of] separating out. Others, like Anaximander and Anaximenes, said that everything was generated directly from a single thing. Others still, such as those who said that the principles were finite, said that generation existed, and made generation [occur] out of one another [of the elements], having observed that one thing's destruction was another's generation.[14]

That the slightest oversight at the beginning appears to 'multiply itself ten thousand-fold' as we go along, he establishes both by induction[15] and by argument. By induction, because when Democritus and everyone else[16] who made such suppositions had posited certain tiny things and smallest magnitudes[17] as principles, since these possessed, as principles, the greatest power, when they made errors concerning them they upset the greatest truths of geometry, namely that magnitudes are divisible to infinity, on account of which it is possible to bisect any given straight line.[18] Thus he elegantly shows that the slightest oversight in the case of the postulation of the smallest magnitude becomes responsible for the greatest errors on account of the very great power it has as a principle.

He establishes the same thing by argument too, by adducing the reason for it.[19] For the starting-point is, he says, greater in potentiality than it is in magnitude; and for this reason an apparently minor oversight concerning the starting-point, being very great in potentiality, becomes manifest as we go along and becomes gigantic. And if the improper assumption of the smallest magnitude in the account of the original principle becomes responsible for the greatest errors, then a badly posited infinite magnitude at the beginning will be responsible for far greater ills. So if the infinite is posited as a principle of things by some people, and is sought in the principles of quantity (for both continuous and discrete quantity are either finite or infinite), it is clear that whether it is infinite or not has the greatest power and will make the largest difference in matters having to do with first principles.[20]

So for this reason we must determine whether the body of the universe is infinite or not. For if there were anything infinite at all, this would be it. And it is clear that the infinite possesses the power of a principle; for he who inquires as to whether the world is infinite or not inquires as to whether there is an infinite principle,[21] since the world could not be infinite if there were no infinity in the principles. But the principles of the world are the simple bodies which are finite in number; so if they are also finite in magnitude, what is [composed] of them will be finite too.

> **271b17-20** Every body must then be either simple [or composite; so an infinite body too must be either simple or composite. Furthermore, it is clear that if the simple bodies are finite, the composite body must be finite too, since what is composed of things finite in number and magnitude is itself finite in number and magnitude; for it is as large as things out of which it is composed. It remains to be seen then whether it is possible for one of the simple bodies to be infinite in magnitude, or whether this is impossible. So let us begin with the primary body, and then consider the others. That any body that moves in a circle must be finite] is clear from the following.

He assumed earlier that if there is an infinite, it must be either simple or compound, on the grounds of its being an infinite *body*, while every body is either simple or composite. He also assumed that what is composed of simple [bodies] which are finite both in number and magnitude must also be finite, because the composite will be of the same magnitude as the things which are used to compose it. Then, having shown that none of the simple [bodies] can be infinite but rather that the simple [bodies] are clearly finite both in number and in magnitude, evidently he has proved that there is no infinite body, whether simple or composite, and consequently that the universe is not infinite. Then turning his attention first to the first of the simple bodies, the heavenly [body], he proves that it must be finite and cannot be infinite.

204,1 **271b28-33** If the body that moves in a circle were infinite, [then the radii from its centre would be infinite. But the interval between infinite lines is itself infinite (by 'an interval between lines' I mean that beyond which it is not possible to assume a magnitude which touches the lines). This, then, must be infinite,] since for finite lines it will always be finite.

That the revolving body is not infinite he shows by the following *reductio ad impossibile*.[22] If the heavenly body were infinite in magnitude, the heaven could not move in a circle. But the heaven does move in a circle. Therefore it is not infinite. The minor premiss,[23] that the heaven moves in a circle, is obvious, and he draws attention to it briefly in saying 'but we see the heavens turning in a circle'.[24]

He proves the conditional[25] as follows: if the heavenly body is infinite in magnitude, the radii from the centre will be infinite in length. But if they are infinite in length, the radii drawn from the centre will have an interval (i.e. the intervening area)[26] which is infinite. And if this is true, this interval will be untraversable, and the heaven will never move around in a circle. That the lines from the centre are infinitely long he assumes to be agreed, since it is evident, if the heaven is assumed to be infinite in magnitude.

That this interval between lines [which are] infinite in length is infinite he proves first by clarifying what sort of interval this is, namely 'that beyond which it is not possible to assume a magnitude which touches the lines'.[27] For it is not possible to take any part beyond this area between the lines which touches them, since the whole is inside, and being inside it touches the encompassing lines.[28] For generally, if the lines are infinite in length, how could there be something which exceeded them and which also touched them?[29] Consequently the interval between lines extended to infinity will not have any partial magnitude outside itself either. For this will exceed the limits of the lines, and they will no longer be infinite.

Therefore if they are infinite, the interval between the lines will not have anything which exceeds it; and what is extended to infinity and has nothing which exceeds it is clearly infinite: for if, while proceeding to infinity it is none the less limited, it would at any rate have something which exceeds it. Thus he rightly made it clear, from the [fact that] there is no magnitude which exceeds it, that the interval between the lines proceeds to infinity, and at the same time demonstrated that the interval between the lines was infinite. For, he says, if the interval between the lines is finite, then they themselves are finite.

However, Alexander says: 'there is nothing of the magnitude' – the interval magnitude – 'which, since it exceeds the lines, touches them, but rather it touches them and does not exceed them'; and this is the interval between them. He too rightly understood how there was nothing of the intermediate interval which exceeded the infinite lines; for if there were some part of it which exceeded the lines, that which was between the lines inside it would have to be finite.[30] The only difference is that Alexander thinks that he [sc. Aristotle] derives the infinity of the intermediate interval not from there being nothing outside it, but solely from the [fact that] the lines from the centre are infinite.

But perhaps Aristotle took it to follow from this[31] that there was nothing which exceeded the interval between the lines from the centre, and from this that the interval between the lines was infinite; for if there is nothing outside it as it proceeds to infinity, this must be infinite. For it is always possible to take something that exceeds something finite, even if it approaches infinity. And it is self-evident that the intervals between infinite lines are infinite from the [fact that] the interval between the lines increases in proportion to the increase of the lines themselves, and so that of finite [lines] will always be finite, as Aristotle himself says. And it follows from this that [the interval] between non-finite [lines] will be infinite.

> **271b33-272a3** Further, if it is possible[32] to take something greater than any given [magnitude], [so that just as in the case of number we call something infinite because there is no maximum,] the same reasoning will also apply to the interval.

This is also indicative of the [fact that] the interval between the lines is infinite, or rather of its increasing towards infinity. For if it is always possible to take something greater in area than any given magnitude in itself, just as there is always something greater than any given number, then it is necessary that, as the number increases towards infinity, so too for this reason the area between the lines will increase towards infinity too. And what increases towards infinity has no limit.[33] Therefore it is infinite and untraversable.

The text is transmitted in two ways, either as 'further, if it is possible', or as 'further it is always possible'.[34] Clearly the former is clearer; and it is also, in my view, unelliptical. For it is clear that the demonstration concerns the infinity of the intervening interval, which follows from there being nothing which exceeds something which proceeds to infinity and from its being possible to take something greater than any given [quantity]. And the 'always' is understood along with 'to take something greater than any given [magnitude]', according to customary geometrical usage.[35]

206,1 **272a3-7** Therefore, if the infinite cannot be traversed, [and it is necessary that the interval of the infinite is itself infinite, it will not be possible to move in a circle.] But we observe the heavens turning in a circle, [and we determine by reason that its motion is circular].[36]

Having shown that the area between two lines radiating from the centre will be infinite if they themselves are infinite, then given that the revolving body *is* infinite he infers that, since the interval between the lines is infinite, it cannot move in a circle. For if the interval between two adjacent straight lines radiating from the centre is untraversable, the entire circumference, of which the interval taken is a small part, will be even more so. So if we observe the heaven turning in a circle every day, it is clear that whichever moving point of the circle is taken, it will traverse in the circuit not just one but all of the intervals between the straight lines radiating from the centre.

However, if the interval between [any] two of the straight lines is infinite, it will not traverse it. So if the heavenly body is infinite in magnitude, the heaven cannot move in a circle. But the heaven evidently revolves in a circle,[37] setting out from and [moving] to the same place, so clearly it is impossible for the heavenly body to be infinite in magnitude.

And having confirmed the minor premiss of the syllogism by perception, in saying 'but we observe the heavens turning in a circle', he says it can be proved by argument as well, drawing our attention to the argument where he proved that there must be some simple body which is of a nature to move with a circular motion in accordance with its own nature.[38]

Alexander, however, proves by a general argument that no body, whether simple or composite, could be infinite. 'For every body', he says, speaking of the simple ones, 'is either dense or rare, since every body falls under these differentiations. And these are either both finite or both infinite, or one of them is infinite and the other finite. But they cannot both be infinite, for then there will be many infinite bodies, which is impossible, since the infinite is everywhere, hence if one were infinite, the other could not be.[39]

Further, what is composed of both of the infinites will either be equal to each of them or greater than each: for it cannot indeed be less.

But if it is greater, infinite will be greater than infinite, and infinite double infinite, if indeed both infinites are equal. But double and half are [found] in finite things.[40] But if that which comes from both of the infinites is equal to each of them, it would be absurd that with an infinite distance added to something no increase occurred, at least if, whenever the smallest body is added to a body, the whole must be greater.' 'Moreover', he says, 'if both simple [bodies] are infinite, either they will be equal or unequal. But if they are equal, they will 207,1
be of equal measures.[41] But things measured by equal measures are finite, and so they are not infinite, and so what is [made] from them will not be infinite either. But if they are unequal, one of them will be greater and exceed [the other]. But everything which is exceeded is, insofar as it is exceeded, finite, and so both of them will not be infinite. And in general the amount by which an infinite exceeds a finite must be either finite or infinite. And if finite, the whole is finite, since that part up to the excess is finite, and the whole is measured out by a finite part. But if the excess is infinite, a part will be equal to the whole, and an infinite greater than an infinite by an infinity, which is absurd, if the whole must be greater than the part.[42]

But if both are finite, then what is [made] from them will also be finite; and if this is so, no body, whether simple or composite, will be infinite. But neither can one of them be said to be infinite, because since it is infinite and everywhere it will not leave over any chink of room for the other, and because, as he himself showed in the third book of the *Physics*,[43] since, because the one of them is superior in the mixture and composition, it would destroy the other. And if this were so, such a thing would not be a composite, and consequently not a world: for the world is a composition of the elements.'

And that the cyclical and spherical body cannot be infinite either, Alexander seeks to prove in the following way.[44] For it is not possible for a circle or a sphere to be infinite, at least if a circle is that figure which is such that there is one single point within it from which all the straight lines drawn to the circumference are equal to one another, and if equality is [found] among finite things.[45] And the same argument applies to the sphere.

Moreover, the non-finite cannot come to an end at some point, while something coming to an end at some point must be finite.[46] But the straight lines come to an end at the circumference. Consequently the interval of both circle and sphere is finite [in length], and hence so too are the circle and the sphere; for they are such as their intervals are.[47]

And in general, to talk of an infinite circle or sphere is to do away with circle and sphere; for they have their being in their being shapes, and no shape is infinite, as their definitions make clear. Moreover, that in which there is extremity and middle is not infinite. Further, that in which there is beginning and end is not infinite; and in these cases, any point taken on the circumference is both beginning and end of the whole.

Again, if those [lines] produced from the centre from which the diameter is [formed] are infinite, then either both will be infinite, or
208,1 both finite, or one of them will be.[48] But if only one is infinite, the [lines] from the centre will not be equal, hence it will not be a circle or a sphere, while if the [lines] from the centre are equal, the finite will be equal to the infinite.[49] But if both are finite the whole diameter will be finite, and so too will be the whole circle and sphere. And if both are infinite, there will be several infinites, and infinite will be greater than, indeed double, infinite.[50]

Moreover, if the diameter is infinite it will have no middle, and if there is no middle of this, there will not be one of the circle and sphere. And if it has no middle it will have no centre, and if no centre, it will not be a circle or a sphere.[51]

But we should realise that, because Aristotle has not yet shown that the heaven is spherical, he did not think it right to employ this demonstration.[52] For if it is agreed that the heaven is spherical, it is superfluous to inquire whether it is finite or infinite; and in general it was first necessary to show that it was finite and in this way to inquire into its shape.

> **272a7-11** And again, if you subtract a finite time from a finite time, [the remainder must be finite and have a starting-point. But if the time of the journey has a starting-point, then there will be a starting-point for the motion, and consequently for the magnitude which is travelled.] And this is equally the case in other contexts.

It was proved in the *Physics*[53] that time is dependent upon motion, and motion is dependent upon the magnitude of the moving thing and that along which the motion takes place;[54] and that if the time is finite the motion too must be finite, and so too must be the magnitude of the moving thing and that through which the motion takes place. For a finite magnitude cannot traverse an infinite distance, nor an infinite magnitude a finite distance, in a finite time, as is there proved.

So if the time of the circular orbit of the heavenly body is determined and finite, and takes place in twenty-four hours, as is the case, then both the motion and the magnitude are finite, and each has a beginning and end, and everywhere the end co-exists with the beginning and the beginning with the end.[55] And if this is so, the revolving body will not be infinite, even if it moves *ad infinitum*.

But although it would be possible to demonstrate what is at issue in this way, he passed over this argument as being obvious, and establishes the same thing in a way by *reductio ad impossibile*. So now, having supposed the heaven to be infinite, he concludes that it will cover an infinite distance in a finite time and that the infinite will have a beginning, both of which are impossible, and in the succeeding argument that the infinite will cover a finite distance in a finite time, which is also impossible.

He proves the present [proposition] as follows. He assumed that the time of the heaven's orbit is finite, being twenty-four hours, and he also assumed that whenever you subtract a finite time from a finite 209,1
time, the remainder must also be finite and have a beginning, insofar as the subtraction from the whole took place. For if you subtract the first hour, the beginning of the second hour is the beginning of the remaining time, i.e. of the twenty-three hours.

And so, taking the example of a journey, he says that if the time of the journey, even if it is continuous, has a beginning in respect of which something is subtracted, that time of the journey before it comes to rest (if there is any) will be finite, and there will be a beginning of the motion and of the magnitude of the moving thing. For the beginning of the motion is established in respect of the beginning of the moving magnitude and of that through which the motion [takes place].

And the case is similar, he says, in other instances of motions, of moving things, and of time measuring the motions. So if the time of a single revolution of the heaven has a beginning in respect of which subtraction from it takes place, then both the motion and the moved magnitude will have a beginning, and will not be infinite. For that which has a beginning also has an end, and is not infinite.[56]

And I think we will learn from his geometrical exposition the reason why, although it is evident that the time of the heaven's revolution is finite, he did not argue in terms of the whole time, but rather [in terms of] that after the subtraction, as something which has a beginning and is finite.

> **272a11-20** So let there be a line, namely ACE, [infinite in one direction, to wit E; and let there be a line BB infinite in both directions. If, then, ACE describes a circle around the centre A,[57] ACE will cut BB in its rotation for a finite time; for the whole time in which the heavens rotate is finite, and therefore so must be the subtracted part in which it cuts BB in its motion. Therefore there must be some starting-point at which ACE first cuts BB: but this is impossible. Therefore it is not possible for the infinite to turn in a circle;] and consequently neither could the world, if it were infinite.

He inscribes two lines in the body of the universe, the first, ACE, with A lying at the centre and describing a circle in respect of C in its revolution around the centre,[58] while the heavenly body extends infinitely in the direction of E. Then he inscribes a second, BB, no longer through the centre but outside it, and separated from ACE; and BB is infinite and motionless in both directions. He draws BB outside the centre so that ACE will not coincide with BB all at once as it goes past it (for he could not thus draw the absurd conclusion that it is traversed although it is untraversable), but so that it cuts it for a short period.

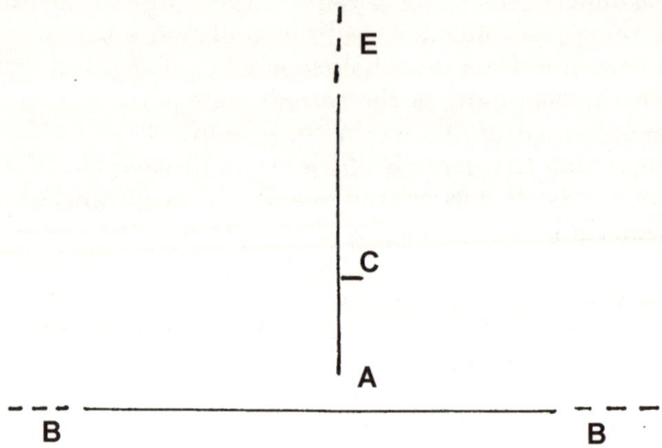

Whenever, then, he says, the [line] ACE of the revolving universe is carried round in a circle so that C describes a circle around the centre A, it is clear that in that revolution there will be some first moment and some first point at which ACE makes contact with BB. For E will not be exceeded by B, since ACE is infinite in the direction of E. And cutting the whole of BB it will pass through it while C returns to its original place, that is while ACE is carried round in a circle. The circle-drawing of C, I believe, is adduced as an indication of this.[59]

So if it is impossible for it to pass through BB, since it is infinite and untraversable,[60] then it is impossible too for ACE, being infinite itself as well, to be carried round in a circle such that C describes a circle. Consequently neither can the world, of which ACE is a [line] from the centre, if it is infinite, turn in a circle, and particularly in the finite time from ACE's beginning to cut BB until it finishes doing so. For I think that it was for this reason that he subtracted the finite time before ACE begins to cut BB from the finite time of the revolution, so that he might show that this postulated cut had a beginning and an end,[61] even though it was untraversable by way of C's describing a circle, through which he showed that ACE[62] returned to its original place.

And even if line BB is not assumed to be infinite, but only a distance,[63] even so ACE will cover an infinite distance in a finite time,[64] which is impossible. And it is clear that the impossibility followed from the [fact that] the time of the heaven's revolution was finite and easily comprehended, given that it takes twenty-four hours, while the moving thing was supposed to be infinite.[65]

But Alexander says that there are two absurd conclusions drawn, both this one, namely that it traverses an infinite distance in a finite time, which he [sc. Aristotle] indicated by 'ACE will cut BB in its rotation for a finite time'; and also furthermore that there will be a beginning for the infinite, which he [sc. Aristotle] makes clear by 'therefore there must be some starting-point at which ACE first cuts BB: but this is impossible'.[66] And it is clear that that absurdity follows from this one, namely that an infinite magnitude such as ACE should cover an infinite distance in a finite time. For he showed this to be impossible in the *Physics*.[67]

But perhaps Aristotle did not now overlook this impossibility[68] as not being to the point, but because ACE, even if it is infinite in the direction of E, is still a line and is moved as a whole in the revolution and not only part by part: for in this way it would be absurd for an infinite magnitude to move as a whole in a finite time.[69]

It is possible to describe as 'subtracted' both the time before ACE moves up to BB and that after which it begins to cut BB until it returns to the same point it set out from. For this too is a part of the 211,1 time of the whole revolution. And one might rather call this the subtracted part, as Alexander says, since 'in which it cuts BB in its motion' fits it, because in the prior time it is not yet cutting; and this too is a finite time subtracted from a finite, and has a beginning. And if the time has a beginning, so too will the motion and the magnitude, both of the moving thing and of that through which the motion [takes place]. And it will be finite: for this time is subtracted from the finite time of the revolution.

But one should realise, as Alexander himself agrees, that the issue here is not to prove that the world is not infinite, but rather that the encircling body [is not infinite]. Alexander puts it in the following words: 'and when he says "and consequently neither could the world, if it were infinite", he means by "world" the revolving body, which a little earlier he called "the heaven". Consequently, even if he shows that the world is not infinite in magnitude, he shows primarily that the heaven is finite; for the world is finite because the heaven is finite. And these are the many methods of argument deriving from motion and the time of the motion which are appropriate to the heaven, and which are undertaken here, since in general what was at issue was proof in regard to the revolving body.'

272a21-b17 Moreover, it is evident from the following [that the infinite cannot move; for let A be moving past B, something finite past something finite. It is clearly necessary that A will get clear from B at the same moment as B from A, since the extent to which one thing overlaps another is the same as that to which the latter overlaps the former. If, then, both of them are moving in opposite directions, they will clear one another more quickly, while if one moves past the other at rest, they will do so more slowly, if the one passing is moving at the same speed in both cases. But this at least is evident, that it is impossible for an infinite line to traverse it in a finite time; therefore it must take an infinite time (this has been shown previously in the books *On Motion*).[70] And it makes no difference at all whether the finite line is moving past the infinite, or the infinite past it, since whenever one is going past the other, the latter is passing the first, whether it be moving or motionless (except that they will get clear of one another more quickly if both of them are moving). However there is nothing to prevent a moving line sometimes passing a stationary one more quickly than one moving in the opposite direction, if one assumes that both of those moving in opposite directions are doing so slowly, while that which is moving past the stationary line is going much faster than either of them. Therefore it is no impediment to the argument to assume it passing a stationary line, since it is possible for a moving A to pass a moving B more slowly. So if the time which a finite line takes to get clear [of an infinite one] is infinite, that in which an infinite line moves a finite distance must be infinite also. Therefore it is impossible for an infinite line to move at all, since an infinite time must elapse whenever it moves even the smallest distance. But in point of fact the whole heaven turns and revolves in a finite time; consequently it goes round any line within that, such as the finite line AB, in a finite time.] Therefore that which moves in a circle cannot be infinite.

That an impossibility follows for those who say that the revolving body is infinite in magnitude, he sets out to show on the basis of the moving thing, and of that through which the motion [takes place], and of the motion itself, and of the time of the motion, all of which are congruent with each other. First he proved that two impossibilities will follow from the hypothesis that the revolving body is infinite in magnitude, even though we observe it to revolve in a finite time: there will be a beginning for the infinite, and it will cover an infinite distance in a finite time.

Now he shows, on the basis of the same assumptions,[71] that an infinite body will completely pass through any distance in a finite time, something which he proved to be impossible in the *Physics*.[72] For it will be possible for something to pass through both a finite and an infinite [distance] in the same time and at the same speed, given that what is moving at whatever speed can pass through a finite [distance]

in a finite time.[73] So if something infinite in magnitude will take an 212,1
infinite time to pass through even a finite distance (if one supposes
some part of the heavenly distance to be a finite distance), while the
heaven can pass through the whole of its distance in a finite time,
then clearly it will not be infinite in magnitude.

And it can be proved from this, he says, that in general it is 5
impossible for something infinite to move in a finite time. For if it is
impossible for an infinite body to move either a finite or an infinite
distance in a finite time, it is clear that it is completely impossible for
it to move.[74] So if the heavenly body moves in a finite time, clearly it
is not infinite.

Wishing, then, to show these things clearly, he first assumes two 10
finite lines, the one moving past the other, and shows that whether
both of them are moving in opposite directions or whether one moves
past the other one at rest, [since][75] they get clear of one another at the
same time.[76] For what is separate is separated from what is separate
from it. It is also clear that the time will be finite when both of them
are finite[77] (by passing over this, in my opinion, he introduced much 15
unclarity into the argument). But if they are both moving in opposite
directions, the time of getting clear will be shorter, while if one of
them is at rest while the other moves past it, it will be longer, unless
the one which moves moves faster than when they were moving in
opposite directions. But whatever the case may be, the time will be
finite.

But if the moving line (or moving magnitude: it comes to the same) 20
is assumed to be infinite, or if the moving thing is finite but that which
it moves through is infinite, it will be impossible for it to pass through
it in a finite time. For in the case of a finite magnitude and a finite
distance, whether they both move in opposite directions or whether
one is at rest and the remaining one moves past it, they will both get
clear of one another in a finite time. And it is equally the case that,
whether a finite magnitude passes through an infinite [distance], or 25
an infinite through a finite, and whether they both move in opposite
directions or whether one moves past the other at rest, the time for
them to get clear of one another will be infinite: i.e. they will never
get clear of one another.

For even if the whole infinite magnitude passes through the small-
est distance or the smallest magnitude through an infinite distance,
the time must be infinite. For in general, if the infinite passes through 30
the finite, the infinite will be measured out by the finite, and will no
longer be infinite. So if the whole heaven, in the finite time of its
revolution, goes around some line which is assumed to be within it,[78]
clearly it is not infinite in magnitude, for then the whole would never
get clear of it [i.e. this line].

Think of a line AB as fixed to the depth of the heaven; and it must 213,1
be conceived as stationary. In the passage where he says 'so if the time

which a finite line takes to get clear is infinite', we need to understand in addition '[in passing] through the infinite [distance]', since he immediately gives the corresponding case in the line 'so also must that in which an infinite line moves a finite distance'. He sets the argument up with lines on account of their abstract clarity, assuming lines in place of the moving thing and that through which it moves.

> **272b17-24** Furthermore, just as a line of which there is a limit [cannot be infinite, or rather if it is it is so in length only,[79] so too a plane cannot be [infinite] in the direction in which it is finite; and if it be determinate, it cannot [be infinite] in any way; that is, an infinite square, or circle, or sphere is as impossible as an infinite length of one foot. So, if neither a sphere nor a circle can be infinite, and given that where there is no circle there is no circular movement, and similarly where there is no infinite circle there is no infinite movement, then there cannot be an infinite body which moves in a circle,] at any rate if there is no infinite circle

Having shown, on the grounds that the motion of the revolving body is finite and that it moves in a finite time, that it is impossible for it to be infinite, he now shows, on the basis of the motion, that the circular motion is not infinite in the sense that it is always coming to be and that the thing moving in a circle never completes its entire [motion], but that it is infinite in that it does so again and again. But in this way the thing that moves in a circle would not be infinite, since [if it were][80] it would not move with a finite motion. This he proves as follows.

Just as a finite line, insofar as it is finite, is not infinite, or if it is it is so only in respect of that part wherein it has length without limit,[81] equally planes, insofar as they are limited, cannot be infinite, even though they may sometimes be infinite in respect of another part.[82] And so too those which are shapes, be they plane or solid, are limited and determined in all directions, because a shape is that which is enclosed by one or more boundaries, and is infinite in no direction. Moreover, a line which is limited at both ends, like a square, is infinite in none.

So if neither sphere nor circle is infinite, and circular motion is given its form by the circle (since if there is no circle there will be no circular motion), then it is clear that if there is no infinite circle there will be no infinite circular motion. And if the motion is limited in distance, the moving thing will be limited in magnitude, since if it were infinite it could not move with a finite [motion]. So just as earlier the finitude of the moving thing was proved from the finitude of the time, so now it will be shown to be finite from the finitude of the motion. And clearly such a proof has proceeded on the assumption of the circularity of the motion itself, since the sphericity of the fifth body has not yet been proved.[83]

At the beginning of the passage some [manuscripts] write 'or rather if it is it is so in one direction only' in place of 'or rather if it is it is so in length only', thus making the thought clearer. 214,1

> **272b25-8** Moreover, if C is the centre, [while line AB is infinite, and line E is infinite at right angles to it, and CD is moving, it will never get clear of E, but it will always be such as CE, since F cuts it.][84] Therefore an infinite body cannot revolve in a circle.

This fifth argument also proves on the basis of the motion that the revolving body is not infinite in magnitude, since if it were infinite, it could not go round in a circle; but it does go round; therefore it is not infinite. The demonstration is as follows.

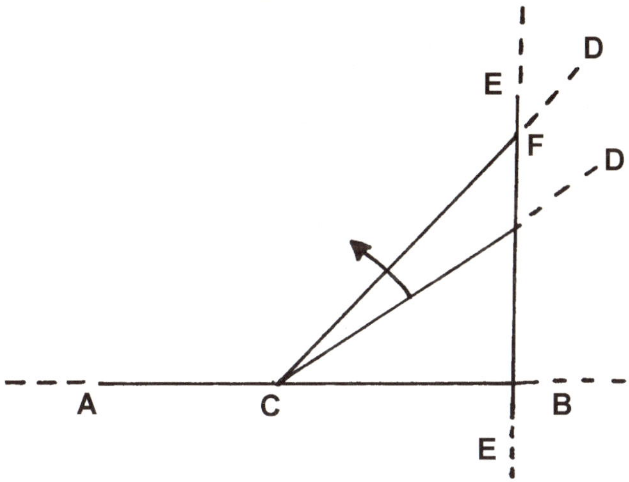

Letting C be the centre of the revolving body, he draws a line AB as a diameter through it infinite in both directions; then he draws a line E at right-angles to it displaced from the centre, also infinite in both directions, and then produces a line CD infinitely from the centre C, cutting E at F; now he says that if all the rest remains stationary, but CD moves with the whole of the revolving body, it will never get clear of E, but will always be such as CE, (i.e. AB).[85]

For the part of it from C is in contact with E. Moreover, the part of CD to F always cuts E at some point, and will never get clear of it because of their both being infinite. And if CD never gets clear of E it will never complete the circuit, and hence neither will the whole body in which CD is [embedded] go round in a circle, given that it is infinite so that both CD and E are produced in it infinitely. But the heaven does go round in a circle; therefore it is not infinite in magnitude.

The expression 'since F cuts it' is unclear, since he refers to CD as F because CD is drawn as cutting E at F, just as he calls AB CE, since AB cuts E at that point.[86]

272b28-273a1 Again, if the heavens are infinite [and move in a circle, they will have traversed an infinite distance in a finite time. For let there be a stationary heaven, and one of the same magnitude moving within it. Consequently if, being infinite, it completes a revolution, it will have traversed an infinite equal to itself in a finite time:] but this is impossible.

Having shown that the revolving body is not infinite on the grounds
215,1 that it would not be able to go round in a circle if it were infinite, i.e. on the basis of its motion, now, on the supposition that it does go round in a circle, he concludes that an infinite passes through in a finite time, which is impossible.

He shows this clearly, but not, I think, as Alexander says he does by positing two spheres, with one of them outside and encompassing the other; for even if the convex surface of the encompassing [sphere] is equal to the concave [surface] of the enclosed one, still the whole heaven will not be equal to the whole, as Alexander says it is. But rather, I believe, he cleverly posits a double distance, the one of the magnitude of the moving thing, the other of the space in which the moving thing moves, the latter being at rest. For these are equal.[87]

And in general, nowhere in this discussion does Aristotle talk of the sphere of the heavens, since he has not yet demonstrated anything about its shape.[88] So he says that if the heaven goes round in a circle, it will pass through an infinite stationary distance in a finite time of revolution, which was proved to be impossible. And this is the demonstration from time.

And it should be noted that it is not only in the case of the sphere that the space is equal to the thing which moves in a circle in it, but in all circular cases, such as the cylinder or the cone. For things with angles, like a cube, need more space when they revolve on account of the protuberance of their angles.[89]

273a1-6 One can put it conversely, [namely that if the time in which it rotates is finite, then it is necessary that the magnitude which it goes around[90] be also finite. But it traverses a distance equal to itself: therefore it too is finite. So it is evident that that which moves in a circle can be neither without limit nor infinite,] but it too has a limit.

On the supposition that the heaven is infinite in magnitude, he proved that it could not turn around in a finite time; for, being infinite, it must go through something infinite in magnitude, which is impossible to do in a finite time.[91]

Now, having assumed that the heaven turns around in a finite time, he concludes that it must itself be finite in magnitude, which is what he means by 'conversely'. For, on the same assumptions as before (I mean that both the moving thing and the distance in which it moves are equal), it is necessary that if the time is finite so too must be the magnitude of that which goes round. But in the case of things moving in a circle, the moving thing and the distance which it covers are equal. Therefore the heaven itself is limited in magnitude.

And it should be noted here that he was not earlier assuming as equal to the heavenly magnitude another sphere,[92] but rather the distance which it goes round, which he also refers to here when he says 'the magnitude which it goes around'.[93]

[CHAPTER 6]

> **273a7-8** What is more, neither that which travels towards the centre nor that which travels away from it will be infinite.

Having concluded that the revolving body is not infinite in magnitude, he goes on to show that none of the other simple bodies is infinite in magnitude either. These are the ones which move in a straight line, some away from and others towards the middle.

It should be noted that even if this[94] is entailed by the demonstration that the whole cosmos is limited in magnitude, given that its parts are limited both in number and in magnitude, even so the primary aim of this chapter is to show that the simple bodies in the cosmos are limited in magnitude just as previously [it was to show] that they were limited in number, when he concluded by saying 'it is evident from what has been said why it is impossible for there to be a greater number of simple bodies than those mentioned'.[95] And there he added the reason why there are no more than three (or five, if one divides the upward and downward ones).[96]

> **273a8-15** For upwards and downwards are contrary movements, [and contrary movements are towards contrary places. And if one of two contraries is determinate, the other will be determinate as well. But the centre is determinate, since no matter where whatever sinks comes from, it is not possible for it to travel further than the centre. So given that the centre is determinate, the upper place must be determinate too. But if the places which are determinate are also finite,] then the bodies too will be finite.

The first argument, which shows that the simple bodies which move in straight lines are finite as well, takes its start from motion and the contrariety in respect of it. Things moving upwards and downwards move with contrary motions. Things moving with contrary motions move from and to contrary places. Things moving to contrary places move to determinate [places]. Things moving to determinate places move to finite [places]. Things moving to finite places are themselves finite, since it is not possible for something infinite to be in a finite place. Consequently fire and earth, of which one tends upward and the other downward,[97] are finite.

All of these premisses are evident; and that which says that the contrary places are determinate is no less clear than the others, if indeed contraries are furthest removed from one another,[98] and the furthest is determined.[99]

But passing over the general demonstration of this, appropriately to the matter in hand, he shows that the middle is determinate from the [fact that] each of the contraries is. And he shows the latter from the [fact that] things moving from everywhere to the middle move towards a determinate middle and do not overshoot. And if the middle is determinate, the upper [place] must be determined too. For if it were indeterminate, it would not be upper, since the upper is a limit in relation to the middle, and nor would it be contrary [to the middle], since it would not be furthest distant [from it]: for the furthest is determined.[100] And if the upper [place] were not determinate but infinite, there would always be a distance further removed from the downmost point than that of any particular intervening space one cares to assume, and there would be no furthest.

Alexander comments that the impossibility of what moves downwards naturally going further than the middle [is evident] both on the basis of perception (for it moves towards the earth) and from the [fact that] if something were to overshoot the middle, at the moment it overshot it it would no longer be moving downwards but towards the upper [place], which is impossible, for it would be contrary to its nature. But perhaps it is a clear demonstration of the [fact that] all heavy mobile objects are orientated towards the centre that they move at right angles [sc. to the surface of the earth].[101]

273a15-21 Furthermore, if up and down are determinate, [the intervening space must be determinate also, since if it were not, motion would be infinite, and it has already been shown that this is impossible. Therefore the intervening space is determinate as well, and consequently so too is the body which either is in it, or can come to be in it.[102] Moreover, bodies which move upwards and downwards can come to be in it, since it is the

> nature of one of them to move away from the centre] and of the other [to move] towards the centre.

Since one place is up, another down, and a third intermediate between them, having first shown that the upper and lower elements, i.e. fire and earth, are determined and not infinite in magnitude on the grounds that upwards and downwards are determined, then having shown the intervening space to be determinate he concludes from this that the intermediate elements, i.e. water and air, are determined and finite in magnitude.

He shows that the intervening space is determined by the following *reductio ad impossibile*. If up and down are determined, the intervening space must also be determined. For if it is not, but the intervening space is infinite, motion from top to bottom and from bottom to top would both be infinite, and the interval would be infinite too, and so the moving things would move an infinite distance in a finite time (which was shown earlier to be impossible).[103] Thus he showed that the intervening space is determinate.[104]

On this basis he argues that 'therefore the intervening space is determinate as well, and consequently so too is the body which either is in it, or can come to be in it', that which is in it being the ensembles of the intermediate elements, while that which can come to be in the intervening space is both that which moves from below upwards, i.e. fire, and that which moves from above downwards, i.e. earth.

For if the one moves upwards and the other downwards by nature, clearly they come to be in the intervening space, and the intervening space will not be infinite. For if it were infinite, none of the upwardly mobile bodies would ever come to be at the top after passing through the intervening space, nor would any of those which move naturally downwards ever come to be at the bottom.[105] But if it is impossible for them to come to be in these places, they will never move naturally 218,1 towards them, since what cannot have come to be is not even beginning to come to be, as he showed at the end of the *Physics*.[106]

He assumes as agreed that some of both the things which move upwards from below and those which move downward from above[107] can be in the middle; for in these cases too we must understand 'have come to be there'.[108] For as this is clearly the case, it is impossible for motions to be infinite. And if the motions are not infinite, then the intervening space will not be infinite either, and if the space is finite, the body in it must also be finite.

Alexander says that the phrase 'and it has already been shown that this is impossible' might refer to the proof[109] which showed that it was impossible for the space within a body which moves in a circle and around the middle to be infinite, because the radii would be infinite and something would have to pass through an infinity in a finite time.

Alexander adduces a similar argument [to show] that none of the bodies which move upwards and downwards are infinite. For if the upper and lower places are contraries, and contraries are the furthest removed from one another, while that of which it is possible to take the greatest amount cannot be infinite,[110] the universe, in which there is an up, a down, and an intervening space, cannot be infinite.[111] And if the universe is not infinite, none of them [sc. the elements] is infinite either, since the parts of something finite are not infinite. And if the places are not infinite, neither will the bodies in the places be infinite. And in general, if there is an intervening space which separates the upper and the lower [places], then what separates the upper from the lower [place] is the furthest [possible distance], and the furthest is determined;[112] therefore the intervening space is determined.

> **273a21-2** And so it is evident from these things that it is not possible for a body to be infinite.

Having shown that both the revolving body and the four simple sublunary bodies are finite in magnitude, he reasonably concludes that no body can be infinite. For no simple [body is infinite], since there is none other besides these five; and no composite [body is infinite], since if the simple bodies are finite in number and magnitude, the composite bodies must be finite as well.

219,1 > **273a22-b15** In addition if its weight is not infinite, [then none of these bodies will be infinite, since the weight of an infinite body must be infinite too. (The same argument will hold for lightness as well, since if there is an infinite heaviness there is an infinite lightness, if the body that rises is infinite). And it is clear from the following. Let the weight be finite, and take an infinite body AB, and of weight C. Then subtract a finite magnitude BD from the infinite. Let its weight be E. E will be smaller than C, since the weight of the smaller is smaller. Now let the smaller divide into it as many times as you like, then as the smaller weight is to the greater, so will BD be to BF, since it is possible to subtract as much as you like from the infinite. If, then, magnitude is proportional to weight, then just as the smaller weight is to the smaller magnitude, the greater will be to the greater. Therefore the weight of the finite and the infinite will be equal. Further, if the weight of the greater body is greater, then the weight of GB will be greater than that of FB, and consequently the weight of the finite body will be greater than that of the infinite. And the weight of unequal magnitudes will be the same, since the infinite is unequal to the finite. It makes no difference whether the weights are commensurable or

> incommensurable, since even if they are incommensurable, the same argument will apply. Suppose for example that if E is measured three times it exceeds the weight C; then three times the weight of the whole of BD will be greater than the weight of C;] consequently the same impossibility will arise.

Having shown on the basis of both motion and places that none of the sublunary elements is infinite in magnitude, he now proves the same thing on the basis of the inclinations, by which I mean both heaviness and lightness, couching his demonstration here too in the form of denying the consequent,[113] as follows.

If one of the sublunary bodies is infinite in magnitude it must be either heavy or light, since the sublunary bodies are of such a kind; consequently the weight of the infinite heavy body will be infinite, and the lightness of the infinite light body infinite. But it is impossible for either heaviness or lightness to be infinite; hence none of the bodies which have weight or lightness (or to put it generally, those which travel in straight lines)[114] is infinite.

He proves the conditional[115] which states that if there is an infinite body which is either heavy or light, its weight or lightness must be infinite, on the basis of the [fact that] it must be either finite or infinite; so, having shown that it cannot be finite it holds of necessity that it be infinite.[116]

Next he proves the minor premiss, which states that it is impossible for weight or lightness to be infinite. He shows that the weight (or lightness) of an infinite body (if there is one) is not finite but infinite by way of an impossible case[117] in the case of the elements. For if someone were to say, he says, that the weight or lightness of the infinite body were not infinite but finite, the weight of an infinite body and of a finite one will turn out to be equal, which is impossible.

He proves that this follows for those who say that the weight or lightness of the infinite body is not infinite but finite by positing an infinite body AB[118] and a finite weight C for it, then subtracting from the infinite magnitude a finite magnitude BD, and assigning a weight E to this which is clearly less than the weight C (since that was of an infinite magnitude, and the weight of a smaller thing is smaller).

On these assumptions, the weight E will either measure out the weight C so as to fit it exactly,[119] or it will not measure it out. Assuming the first [case], he makes it measure it out, and as the lesser weight is to the greater weight, so too is the magnitude BD (which is that of the lesser weight) in relation to some other greater magnitude BF subtracted from the infinite. Consequently C is the weight of the magnitude BF just as E was of BD. But C was also the weight of the infinite body AB; therefore both an infinite and a finite magnitude will have the same weight (he himself says 'equal', since the same quantity is necessarily equal): but this is impossible.[120]

But then he draws a consequence even more impossible than this one. For if the weight of a greater body is greater (if the body is uniform),[121] there can be taken a magnitude BG greater than the magnitude BF,[122] since it is possible to subtract something larger from the infinite than any proposed finite [magnitude]; and the weight of BG will be greater than the weight of BF. But the weight of BF was C, which was also [the weight] of the infinite, consequently the weight of the finite BG will be greater than that of the infinite, which is even more impossible than that it be equal to it.

Moreover, he draws a further third absurd consequence, although one which is looser[123] than the previous ones, that unequal magnitudes have the same weight: for the infinite is unequal to the finite.

But since he arrived at the aforementioned conclusions by supposing that the weight E was commensurable to the weight C, he says it makes no difference whether the weights are commensurable or incommensurable,[124] since the very same absurdities will also follow if they are supposed to be incommensurable. For if the weight E of the body BD, when measured three times against C, exceeds it: then whenever we multiply the magnitude BD by three, the weight of the resulting [body], composed of three weights E, will exceed the weight C, and so once again the weight of a finite [body] will be greater than that of an infinite (for C was [the weight] of an infinite [body]).[125]

And if E multiplied by three only fails to exceed C by a part of itself, when multiplied by four it will exceed it, and the same things will be said. Consequently, whether one supposes the weights to be commensurable or incommensurable, the same impossibilities will follow for someone who says that the weight or lightness of an infinite [body] is finite.

> **273b15-26** Furthermore, we can take commensurable weights [(for starting from the weight or from the magnitude makes no difference); for example: suppose one takes a weight C which is commensurable with E, and subtracts from the infinite a magnitude (call it BD) which has a weight E, then just as weight is to weight, so BD will be to some other magnitude (call it BF); for since the magnitude is infinite, it is possible to subtract as much as one likes. On these assumptions, both the magnitudes and the weights will be commensurable with one another. Nor indeed will it make any difference to the demonstration whether the magnitude is uniform or non-uniform in regard to weight, since it will always be possible to take bodies of a weight equal to BD from the infinite] by either subtracting or adding *ad libitum*.

He rightly notes that there is no need of this subdivision, which divides the supposition into [one involving] commensurables and [another involving] incommensurables. For it makes no difference whether we start from the weight or the magnitude. Whenever we start from the weight, and take some weight smaller than but commensurable with the finite weight C (which is [the weight] of the infinite), such as one a third [the size] of it, namely E, and then subtract some magnitude BD, whose weight is E, from the infinite magnitude, then it will come about that just as the weight E stands in relation to C, so too will the magnitude BD stand in relation to the [finite] magnitude BF. For one can subtract from the infinite as much as one likes.

If things are thus, both the magnitudes and the weights will be commensurable and the impossibilities that derive from their being commensurable will follow, and there will be need for the supposition of their being incommensurable. For the weight of BF will be equal to the weight C, which is that of the infinite. And even if we take something larger than BF from the infinite magnitude, its weight, which is the weight of that finite [magnitude], will be greater than the weight C, which was the weight of an infinite magnitude.

But since of bodies some are uniform and others non-uniform,[126] which is the same as saying [they are] of uniform or non-uniform weight, it will make no difference, he says, to the demonstration whether we assume that the infinite is uniform or non-uniform, since in the case of the non-uniform bodies, the proportion will be determined not in respect of magnitude but in respect of weight.[127]

For we will no longer simply subtract from the infinite things equal in whatever way to BD, but rather things equal in weight. For we are employing the proportion in weight, and it is possible to subtract from the infinite in respect of magnitude whatever magnitudes we like equal in weight, sometimes subtracting from the weight of the things taken, and taking things smaller than those previously taken (whenever those parts of the infinite happen to be heavier in constitution), and sometimes adding to the magnitude of the things subtracted, and making them bigger than the ones previously subtracted (whenever those parts of the infinite have a less heavy constitution).[128]

But perhaps one should note, however, that whenever the weight of the infinite [body] is supposed to be finite, even if the infinite magnitude is not exhausted by subtraction, the finite weight will be. And he shows that one may conclude that the weight of the finite magnitude, i.e. BG, is greater than that of the infinite, which is C. But if this is the case, there will be an infinite magnitude as a remainder which has no weight, which is even more impossible.[129]

273b26-30 Consequently it is clear from what has been said [that there will be no infinite body of finite weight. Therefore it will be infinite. So if this is impossible, it will also be impossible for there to be an infinite body. But that it is impossible for there to be an infinite weight] is evident from the following.

He draws the conclusion from what has just been demonstrated, that the weight (or the lightness: for the same argument applies in both cases)[130] of an infinite body cannot be finite. So if bodies must have either weight or lightness (for the argument here concerns sublunary bodies), then whenever some infinite body is posited, it must possess either finite or infinite weight or lightness; and it has been shown that
222,1 it is impossible for them to be finite: so they must therefore be infinite. So if in general it has been shown that infinite weight or lightness is impossible, it is clearly also impossible for there to be an infinite body. So in addition to showing that an infinite weight is impossible, he supplies [the same conclusion], and clearly on the basis of the same considerations, in the case of lightness as well.[131]

> **273b30-274a3** For if a certain weight moves a certain distance [in a certain time, then one so much and more will do so in that much shorter a time, and whatever the ratio of the weights, the times will vary inversely: thus for example if half the weight does it in this time, its double will do so in half the time. Again, a finite weight will traverse any finite distance] in some finite time.

Proposing to show that there cannot be an infinite weight, he first assumes a certain three [propositions] as axioms relevant to the demonstration.

First, that where one of two weights is smaller and the other greater – which is what 'one so much and more' means: for the greater has the lesser [weight] and something more still in addition to it – so the greater of these weights will move the same distance in a shorter time than the smaller.[132]

Second, he assumes something as a corollary of the first,[133] namely that the weights and the times stand in inverse proportion. For if the greater weight is double the smaller, the time in which the greater moves a certain distance will be half that in which the smaller moves the same distance. So if, when the time is half the weight is double, while when the weight is half the time is double, the times are reasonably said to be in inverse proportion to the weights.[134]

The third initial assumption is that a finite weight will move any finite distance in a finite time: for he showed in the *Physics*[135] that nothing finite traverses a finite distance in an infinite time. He employs this to show that every time, even that in which the smallest weight traverses the greatest distance, and that in which something many times heavier than it covers the same distance, have a ratio to one another.[136] For if the finite weight traversed the finite distance in an infinite time, there would be no ratio or proportion of the times to one another which he could use in the proof.

It is clear that every finite weight traverses every finite distance in a finite time; for since a finite weight will traverse a part of the distance in a finite time, no matter how much bigger the whole distance is than the part, it will take that much more time to cover the whole distance than [it took to cover] the part.[137] And he makes frequent use here of what was demonstrated in the *Physics*,[138] namely the impossibility of traversing an infinity in a finite time.

> **274a3-13** Therefore it is necessary from this [that if there is an infinite weight it must both move (in view of the fact that it is as great as the finite and more), and yet also not move (in view of the fact that it must move in proportion to the excess, although inversely, the greater the weight, the shorter the time). But there is no ratio of the infinite to the finite, while there is of lesser time to the greater finite weight. But perhaps it always does so in a shorter time? But there is no shortest time; and even if there were, it would be of no help, since a certain other finite weight has been taken[139] that stands in the same ratio to another greater one as [it does] to that [time] in which the infinite [moves],[140] so that the infinite would move the same distance in the same time as the finite.] But that is impossible.

223,1 Relying on what has been said, he shows by way of *reductio ad impossibile* that there is no infinite weight. For if there were, it will follow from the assumptions that the same thing will both move and not move, which is impossible. For it must move, on account of the first of the things taken to be axiomatic,[141] for if a certain weight moves a certain distance in a certain time, and 'so much and more'[142] moves the same distance in a shorter time, then since the infinite is 'so much and more', clearly it will move in a shorter time.

But on the other hand it cannot move, because it was established that whatever proportion the weights which move the distance have in relation to one another, the times in which each of them moves have that proportion inversely to one another, and that of the times is clearly the inverse of that of the weights.[143]

And each finite time has a ratio to every other. So if the time in which a weight which is 'so much and more' (i.e. an infinite [weight]) [moves], being less than the other time, is still finite, it is clear that the ratio of time to time will be of smaller to greater, finite to finite, while there will be no ratio of the weight which is 'so much and more' (i.e. the infinite [weight]) to any finite weight. But the same proportion must hold inversely both for the weights moving the same distance and for the times. Consequently there will be no time in which an infinite weight will move a finite distance; so it will not move.[144] For every mover has been shown to move in time.[145]

And even if one does not assume this inverse proportion [to hold] between weights and times, but only that the times are finite, some ratio will thus hold for them too.[146]

He added 'but perhaps it always does so in a shorter time?' to show that even if the time becomes shorter, there will still always be a proportion in the shorter time,[147] and there is no shortest time, since it is something continuous on account of its infinite divisibility.[148]

And, he says, if there were a shortest time,[149] it would be of no help to those who think that the infinite weight can move in it, and who thus seek to evade the refutation by way of proportion. By assuming such a shortest time and an infinite weight that moves in it, he derives the following impossibility, namely that the infinite weight moves the same distance in the same time as a finite one.

Let an infinite weight have moved, in the shortest time, a certain distance. Clearly this will be finite, since nothing can move an infinite distance in a finite time, particularly in the shortest time, whether the mover be finite or infinite. For in what time will it move part of the infinite distance? There is nothing smaller than the smallest.

Since, then, it was established in *On Motion*[150] that every finite weight will traverse every finite distance in a finite time, it is clear that some finite weight will traverse this finite distance, which the infinite moves in the shortest time, in a finite time.[151] So, as the distance is the same, just as the times stand to one another, both the shortest in which the infinite weight moves and the finite time in which the finite [weight moves], so too will the moving weights stand to one another, which is impossible. For there is no ratio of the infinite to the finite.

He himself does not infer this absurdity, but rather that the infinite [weight] will move an equal distance in the same time as a finite one. Thus he says that not even the supposition of a shortest time in which the infinite weight moves the finite distance escapes the absurdity. For some other finite magnitude 'has been taken' (in place of 'might be taken')[152] that moves the same distance in a finite time, being in the same ratio towards something else greater in weight than itself[153] as the shortest time in which the infinite weight moved was to the time greater than the shortest in which the lesser of the finite weights moved:[154] for weights are the inverse of the times, as has been said.[155] Therefore both the infinite weight and the greater of the two finite [weights] taken will move in an equal time, or rather in the same very shortest time.

Or did he mean 'has been taken' more precisely?[156] For if the infinite is supposed to move in the shortest time, to this very thing some other finite weight has been taken, standing in the same ratio to another greater one as that in which the infinite [stands to it].[157]

For if the greater moved in the shortest time (because as the greater time is to the shortest, so too is the lesser of the finite weights which moves in the greater time to the greater [weight which moves] in the shortest [time]), and the infinite weight moves also in the shortest time, it is clear that the lesser of the finite [weights] will stand in the same ratio both to the greater and to the infinite. And this too is in itself absurd, that there is a ratio between finite and infinite. But he infers the other [absurdity], namely that both finite and infinite move an equal distance in the same time.

It is possible, Alexander says, in the case of the finite distance, taking its parts, to say that the infinite weight will cover them in less than the shortest time.[158] But nothing is smaller than the shortest. Consequently, either there is no shortest time, or, if there is, in the shortest time [it moves] the shortest distance, which is clearly the same:[159] for there is only one shortest. But the infinite weight and the finite, indeed every unequal weight, will move [the shortest distance in the shortest time], which is particularly absurd. For thus things differing enormously from one another in force will move at the same speed, and the infinite heaviness [at the same speed] as the finite. For the [distance] which the lesser weight has moved in the shortest time, this [distance] the heavier either will not move, or will not move in time,[160] or [will move] in [a time] shorter than the shortest [time]. And if these things are impossible, it too will move in the shortest time.

Alexander says that he constructed this demonstration making use of the third of those things taken earlier to be axiomatic, namely that a finite weight traverses any finite [distance][161] in a finite time. And the passage, he says, is congruent with this: for another finite [weight] has been taken in the same ratio to another greater one as this itself has to the infinite.[162]

> **274a13-17** Furthermore, it is surely necessary [that if an infinite weight can move in whatever finite time, there will be another finite weight that will move a certain finite distance in the same time.] Therefore it is impossible for there to be an infinite weight.

He said as though on the basis of some supposition that 'some other finite weight has been taken'[163] which moved in the shortest time, in which the infinite did as well, at any rate given that one assumed a shortest time. Now he adds that it is also necessary, given that the infinite weight is assumed to move a finite distance either in the shortest or in any finite time whatever, that some other finite weight will also move the finite distance in the same time.

So as the time is the same, it will be the case that just as the finite distance which the finite weight moves in the same time stands to the finite distance which the infinite weight moves, so the finite weight will stand to the infinite weight. And if the weight which has been taken[164] is as many times greater than the finite weight[165] as the distance which the finite weight moves is smaller[166] than that which the infinite [moves] in the same time, then both that[167] and the infinite will move the same distance in the same time.

So if it is impossible for the infinite to have any ratio to the finite, and it is also impossible for unequal things to move the same distance in an equal time, it is impossible for there to be an infinite weight. And substituting any time whatever for the shortest time[168] made the demonstration more general.

274a17-19 The same goes for lightness. [Therefore for bodies to have either infinite weight] or infinite lightness is impossible.

What has been said about weight applies equally in the case of lightness, if we substitute upward motion for downward. So if neither heavinesses nor lightnesses are infinite, neither will the bodies possessing weight or lightness be infinite; for this was already proved.

226,1 **274a19-24** That there can be no infinite body [is clear both from examining each of them individually in this manner and from general considerations, and not only along the lines of the arguments which we rehearsed in *On Principles*[169] (for there we first of all determined in general the circumstances in which there could and could not be an infinite),] but now also in another fashion.

He proved that none of the simple bodies is infinite in magnitude; for the revolving body is not, and nor are any of the sublunary ones, given that the former possesses neither weight nor lightness, while of the sublunary ones the heavy (i.e. earth and water) move towards the bottom while the light (air and fire) [move] towards the top.

The simple bodies had already been shown to be finite in number. For they were shown to be five in all, three if the sublunary ones are reduced to two, the heavy and the light.[170] And having now shown that each of them is finite in magnitude, it is clear that there will be no composite infinite body either. Thus he draws this confident conclusion, saying that it is clear that there is no infinite body, whether simple or composite, on the basis of particular investigations, namely those of each of the simple [bodies].

But since the mode of demonstration by way of the general is more scientific than that by way of particulars, he proposes to show generally as well that there is no body infinite in magnitude, reminding us first of all of the arguments in the third book of the *Physics*[171] which showed generally that there is no infinite body, in order not to say the same things twice.

Then he adds some other things derived from another investigation. And he calls the first four books of the *Physics* 'On Principles', just as a little earlier he called the remaining four 'On Motion',[172] when he said 'but this at least is evident, that it is impossible for an infinite line to traverse it in a finite time; therefore it must take an infinite time (this has been shown previously in the books *On Motion*)'.[173]

> **274a24-8** After this, we should consider [whether even if the whole body is not infinite, none the less it is sufficient for several heavens. For perhaps someone might wonder why anything should prevent there being other worlds than this one, established just as ours is,] even if there is not an infinite number of them.

Having set out to offer a general demonstration of the non-existence of an infinite body, he first mentions something which he will go on to speak of subsequently,[174] after this demonstration.[175] This is that if the whole body is finite, is it sufficient for several heavens (i.e. worlds), or [only] one? For insofar as the whole body is finite, nothing prevents there being other worlds similar to ours (as long as they are not infinite in number: for then the total magnitude would have to be infinite).

> **274a28-9** First let us speak in general terms about the infinite. 227,1

We must first recall the things which were proved generally in the third book of the *Physics*[176] about there being no infinite body, since he himself recalled them, and thus connected them with the arguments here.

He proves there that there is no infinite body, arguing logically[177] that every body is determined by a plane, and what is determined by a plane is not infinite;[178] and physically[179] from the fact that the elements are finite in number and so too finite in magnitude. For all of them cannot be infinite, since there could not be many infinites;[180] but nor could one of them be [infinite]; for if they are to be preserved and generated from one another, their powers must as far as possible be equal and proportional.[181]

So if the power of something infinite is infinite, but there is no ratio of the infinite to the finite, clearly it is not the case that one of them will be infinite. So if both the number and the magnitude of the simple [bodies] is finite, clearly the magnitude of whatever [is made] from them will be finite.

But neither can something infinite in extent underlie the elements from which they are generated,[182] as Anaximander and Anaximenes[183] held, or else there would be a resolution of things which are destroyed into it.[184] Moreover, if there is a proper place for every natural body, towards which it moves and in which it rests, and there is no proper place for an infinite [body], then there will be no natural infinite body.[185]

Furthermore, the infinite will have no more tendency to move downwards rather than upwards, or to remain at rest,[186] since it will be similarly disposed towards all places.[187] And again it is impossible for there to be an infinite number of elements differing from one another in form; for then there would be an infinite number of places differing from one another.[188] Finally [it can be shown] from the [fact that] all the elements move naturally either upwards or downwards, but nothing is either up or down in the infinite.[189]

[CHAPTER 7]

> **274a30-b8** It is necessary that every body be either infinite [or finite, and if infinite then either non-uniform or uniform, and if non-uniform then composed either of a finite or of an infinite number of forms. That it cannot at any rate be composed of an infinite number is evident, if our first suppositions are allowed to stand. For since the primary motions are finite, the forms of the simple bodies must be finite as well. For the motion of a body is simple, and the simple bodies are finite [in number]; and it is invariably necessary that each natural body have a [natural] motion. On the other hand, at least if the infinite is to be composed of a finite [number of forms], each of the parts must be infinite, the water for instance, or the fire. But this is impossible,] since it has been shown that there is neither infinite weight nor infinite lightness.

Having set out now to investigate in general whether the whole body is infinite or finite, he sets out the argument by division,[190] saying 'it is necessary that every body be either infinite or finite', in order to show that, since nothing is infinite, it is finite; for the division is negative.[191]

[To show] that it is not infinite, he makes another division, saying that if it is infinite it is either non-uniform or uniform,[192] since this division is immediate.[193] He sets the argument out in the form of a denial of the consequent.[194] For if it is necessary that if there is an infinite body it is either uniform or non-uniform, and it is shown that neither of these is infinite, then it is clear that in no way can it be infinite.

Seeking first to show that it is not non-uniform, he makes use here too of a division, saying that if it is non-uniform, then it is [composed] either of finite or of infinite forms; so if it is [composed] neither of finite nor infinite forms, clearly it is not non-uniform. But he says he will show that the infinite is not non-uniform and [composed] of infinitely many different forms, if the first suppositions are allowed to stand. By 'the first suppositions' he means those which were initially assumed at the outset. These are that every body possesses an intrinsic natural motion in respect of place,[195] and that there are three simple motions,[196] and that the natural motions of simple bodies are simple,[197] and that there is a single natural simple motion for each of the simple bodies.[198]

If these things hold good, then the forms of the simple bodies must be finite in number. For if all simple natural bodies possess simple motions in respect of place, one for each of them, and the simple natural motions are finite in number (for they are either three or five in all),[199] and the finite motions make for natural movers which are finite in form (since the natural motions provide the form of natural bodies, because nature is a principle of motion for those which possess it),[200] then, employing these prior suppositions, he shows that the simple bodies are either three or five, and neither more nor fewer. For air differs from fire in being light but not the lightest, and water from earth in being heavy but not the heaviest, and thus the three become five. Consequently, if one supposes the infinite body to be non-uniform, the things it is composed of will not be infinite in form.[201]

On the other hand, neither can the parts of the non-uniform infinite be limited in form, since if they are limited in form, necessarily each of the parts must be infinite in magnitude. For if the parts are limited in form and limited in magnitude, then the whole will be limited in magnitude, yet it was supposed to be infinite. So if it is impossible for any of the parts, such as water or fire or the revolving body, to be infinite in magnitude,[202] it is clear that the infinite will not be [composed] of a limited number of forms.

But why, if the non-uniform infinite is [composed] of a limited number of forms, is it necessary that *each* of the parts be infinite? For it is possible that if one of the parts were infinite the whole would be infinite. Or does he leave out this supposition, since it was proved earlier that, if they were not all of an equal magnitude but one of them were infinite, it would destroy the others?[203]

On what basis, then, is it clear that it is impossible for each of them, such as water and fire, to be infinite? Because, he says, it was proved that neither weight nor lightness could be infinite, and because the 229,1
powers of finite bodies are finite, since [the powers] of infinite [bodies] are infinite. So if these things are heavy and light, they cannot be infinite in magnitude.[204] And it was proved that the revolving body is not infinite, and so the whole cannot be infinite.[205]

274b8-22 Furthermore, their places must be infinite in magnitude too, [and consequently the motions of all of them would also be infinite. But this is impossible, if we assume our first suppositions to be true; and it is possible neither for the downwardly mobile to move infinitely, nor for the upwardly mobile according to the same reasoning; for it is impossible that what cannot have come to be should be coming to be, whether in quality, quantity or place. What I mean is, if it is impossible to be white, or a cubit long, or in Egypt, then it is impossible to be coming to be any of these things. Therefore it is impossible even to move towards a place which it is impossible to arrive at by movement. Further, if they are dispersed, it may still be possible for the fire [made] of all of them[206] to be infinite. But body is that which is extended in all directions; consequently how could there be several dissimilar bodies, and yet each of them be infinite?] For each of them would have to be infinite in all directions.

He adduces another demonstration of the same thing,[207] one [derived] from motions and places. For if each of the parts is supposed to be infinite in magnitude, their places must be infinite too.

By employing what had already been demonstrated, he could have shown that neither the upper nor the lower place nor the intervening space was infinite, so that none of the bodies that come to be in them could be infinite in magnitude either. But he omits the demonstration of this since it has already been stated,[208] proving instead that the places are not infinite on the basis of the motions to and through them. For if the upper, lower, and intermediate places were infinite, the motions through the intermediate space of both the things moving downwards from above and those moving upwards from below would be infinite as well, since if the distance is infinite, the motion is infinite too.[209]

So if the first suppositions are true – he now means by 'first suppositions' those proved in the eighth book of the *Physics*[210] – namely that that which cannot have moved somewhere cannot even begin to move towards it, either in quality, quantity, or in respect of place, and in general what cannot have come to be cannot even begin to come to be.[211] So if it is impossible, if the places from, to, and through which motion takes place are infinite, for something moving to come to be either up or down, it will therefore be impossible also for there to be such a motion, and so nothing will even begin to move upwards or downwards.

So if things evidently move, some upwards and some downwards, it is clear that there will not be infinite places; and if not the places, then not what is in the places either. Aristotle takes his argument as far as this, namely that nothing could move either upwards or downwards, and omits the rest as being evident consequences of it.

He wished to add a third argument showing that it is impossible for each of the parts of the non-uniform to be infinite on the basis of the [fact that] if each of them is everywhere it would allow no room for the others. He first shows that [this is the case] even if you do not assume that each one is a single continuous infinite,[212] but is rather put together from scattered pieces infinite in number, either in the way Anaxagoras apparently held, as uniform substances infinite in number,[213] or in the manner of those who hold the worlds to be infinite in number.[214]

For on their account,[215] earths and waters and each of the others are infinite, since each of them is to be found in each cosmos, and consequently out of all the dispersed parts of fire there will be an infinite [quantity of] fire. So if body is that which is extended in all directions, and an infinite body is that which is dispersed infinitely in all directions, how could there be several dissimilar infinites? Perhaps one might think them to be similar on the grounds that everything is one:[216] but how could they be dissimilar? For if the fire is infinite, there will be no place available for any of the other elements to occupy.

Consequently if it is necessary for the parts of a dissimilar infinity (given that they are not infinite in form but, if they exist at all,[217] finite) to be infinite in magnitude,[218] and this has been shown to be impossible, it is clear that the parts of a non-uniform infinity could not be finite in form.[219] But it has been shown that they are not infinite [in form] either.[220] So if it is necessary that the non-uniform infinite, if it exists, either have parts infinite or finite in type, and neither of these is possible, it is clear that there will not be any infinite non-uniform body.

> **274b22-9** But again, neither could the infinite body be wholly uniform. [First of all, since there is no other motion apart from these, it would then have one of them. But if this is so, then it will follow that there will be either infinite weight or infinite lightness. Nor again can the body which moves in a circle be infinite, since it is impossible for the infinite to move in a circle. For this is no different from saying that the heaven is infinite,] and this has been shown to be impossible.

Having said that if there is an infinite body it must be either non-uniform or uniform, and having shown that it cannot be non-uniform, he next demonstrates that it cannot be uniform either, again on the basis of natural motions.

For if the infinite were uniform, the whole would possess a single motion. So since there are three natural motions, upwards, downwards, and in a circle, if it moved upwards and were infinite it would possess infinite lightness, while if [it moved] downward [it would possess] infinite heaviness. So if it has been shown that infinite lightness and heaviness are impossible, the infinite cannot move either upwards or downwards. And generally, if it moved upwards it would be fire and if downwards earth, but it was shown earlier to be impossible for fire and earth (and in general for the light and the heavy) to be infinite in magnitude.[221]

Moreover neither could the revolving body be of such a kind, since it was shown that the heaven is not infinite.[222] So if the uniform infinite must be supposed to possess one and the same motion, and what possesses a single [motion] will be either heavy or light or revolving, and it has been shown that none of these can be infinite, it is clear that the infinite cannot be uniform either.

231,1 **274b29-32** But in fact it is not possible for the infinite to move in any way whatsoever, [since it will either move naturally or by force; and if it moves by force, there will be some other [movement] natural for it, and consequently some other place equal in magnitude to it towards which it will move.] But this is impossible.

Having shown individually in each case that it is not possible for the infinite to exist either as uniform <or as non-uniform>,[223] and hence that it cannot exist at all (given that it must be either uniform or non-uniform), he now shows generally that it is not possible for the infinite to move at all.

For a natural moving body must move either naturally or forcibly, and if forcibly it must at all events possess some natural motion. For we say that things move forcibly and unnaturally whenever they move contrary to their natural motion, since the unnatural and forcible takes second place to the natural.[224] So if it is of a nature to move naturally, and what moves naturally moves towards its own place, there must be some other space equal to it towards which it will move. But this is impossible, since if it were infinite it would already have occupied the whole of space prior to moving.[225]

This argument applies to things moving and changing their place in a straight line, with which the argument is primarily concerned. For what moves in a circle revolves in its own place.[226] And it is worthy of note, I think, that place is here conceived not as the limit of what encompasses,[227] but rather as space and extension; for what could encompass the infinite so that its limit was the place of the infinite beyond the infinite, and how could what was encompassed still be infinite?[228]

274b33-275a14 That it is completely impossible for an infinite [either to be affected in any way by something finite, or to do anything to something finite, is evident from what follows. Let

> there be an infinite body A and a finite body B, and a time C in which it moves or is moved. If, then, A is heated, propelled, or affected in some other way by B, or is moved in any way whatever in the time C, let there be some D smaller than B, and assume that the smaller makes a smaller move in the same time; and let E be the amount altered by D. As D stands to B, so E will stand to some finite quantity. But one may assume that an equal thing will alter an equal amount in an equal time, and a smaller a smaller amount in an equal time, and a greater a greater amount, greater in the same proportion as the greater thing stands to the smaller. Therefore, the infinite will be moved by no finite thing in any time, since a smaller thing will be moved by a smaller in the same time, and anything to which it is proportionate will be finite,] since the infinite stands in no relation at all to the finite.

By means of another argument, he shows generally that none of the natural bodies is infinite, by adopting a further axiom [to the effect] that every natural body either acts, or is affected, or both.[229] Having shown that the infinite neither acts nor is affected, he concludes that there is no natural infinite body. And he shows exhaustively that the infinite neither acts nor is affected by showing that the infinite is neither affected, either by the finite or by the infinite, nor acts, either towards the finite or towards the infinite.

The proof is generated on the basis of the following axiom which was adopted earlier: equal agents produce and move equal amounts in an equal time whenever the movers are equal, while if one of the agents is greater and the other smaller, in the same time the one will produce something greater and the other something smaller, and greater by as much as it is itself greater, and smaller by as much as it is smaller.[230]

So in this case, he says, let there be an infinite [body] A and a finite
one B, and let the time in which the infinite is affected in some way 232,1
by the finite be C, since every motion [occurs] in time.[231] So if the infinite is affected in some way by the finite in time C, whenever we take something smaller than B, the smaller thing will clearly move something smaller in the same time than that which B moves. Then assume something smaller than B, namely D. This, then, in time C will move and produce something smaller than the infinite A, which B moves.

Let E be what is changed by D; for it is clear, as we said, that in the same time the smaller will move something smaller than the greater [will]. So if we make what is moved by D, namely E, stand in the same relation to something else[232] as D the mover stands in to B which itself causes motion and is larger, it will clearly be proportional to the finite. For however much larger B is than D, if we assume something as many times greater than E which is affected by D, it too will be finite.

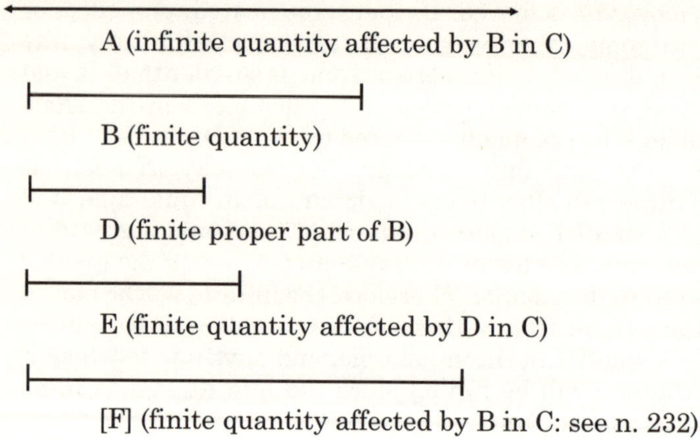

Fig. 1

So whenever the infinite is assumed to be moved by something finite in a certain time, something smaller can be assumed to be moved in the same time by something smaller than B, which is how D moves E. So whenever E stands to something else as D does to B, this other, being finite, will be moved by B, because it stands in a finite ratio to it.

But this is impossible, because the infinite was moved by B, and the infinite and the finite cannot both be moved in the same time by the same thing, since the same thing moves equals in the same time. Nor is it possible for A to stand to B in the same ratio as E to D, so that they can be moved by the same thing, since the infinite stands in no ratio to the finite.[233]

He created an unclarity in the argument by inserting the axioms in the middle, when he says 'but one may assume that an equal thing will alter an equal amount in an equal time', etc., and thus inferring the conclusion by way of 'therefore, the infinite will be moved by no finite thing in any time'.

And he adduced the reason in an unclear manner: 'since a smaller thing' – i.e. E – 'in the same time' – C – 'will be moved by a smaller' – D – 'and anything' – smaller, i.e. E – 'to which it is proportionate' – whenever it happens that as D is to B, so E is to something else – 'will 233,1 be finite'. For this will be finite, and will be in the same ratio to B as was the infinite which was moved by B. So if this follows he might reasonably say that 'the infinite will be moved by no finite thing'; and he adduces the reason for this at the end, saying 'since the infinite has no ratio to the finite'.

Hence the whole structure of the argument would be as follows: if the infinite moves or is affected by the finite, there will be a ratio of the infinite to the finite, and something finite and something infinite will both be moved by the same thing in the same time. But these things are impossible; so by the second indemonstrable[234] the infinite will not be affected by the finite.

> **275a14-24** On the other hand, neither can the infinite move the finite in any time. [Let A be infinite, B finite, and the time in which, C. In the time C, D will move something smaller than B: let it be F. Then let E stand in the same relation to D as the whole of BF does to F. Therefore E will move BF in C. Thus a finite and an infinite will cause [the same] change in the same time. But that is impossible, since it is assumed that the greater does so in a smaller time. But whatever time is assumed it will do the same, so there will be no time in which it will move. But nor yet can it move or be moved in an infinite time, since then it will have no limit,] while action and affection have limits.

Having shown that the infinite is not affected in any way by the finite, he now shows that the finite cannot be affected by the infinite either. And he concludes this by way of *reductio ad impossibile.* For in this case, the infinite and the finite will both alter the same thing in the same time, which is impossible. For it was established that the greater alters [something] in the smaller time.[235] Let some finite magnitude BF be moved in time C by an infinite power A, and let some finite capacity D be assumed. This will then move something less than BF in the time C, since, being finite, it is less than the infinite A.

Then let what is moved by D in the time C and is less than BF be F. If the magnitude BF is finite, and F is finite, some ratio will hold between them. Then let power D stand in the same ratio to some other finite power as F does to BF, and let that be E. So as F is to BF so is power D to E, [and as D is to E, so is F to BF];[236] by alternation, therefore, as D is to F, so E is to BF.[237]

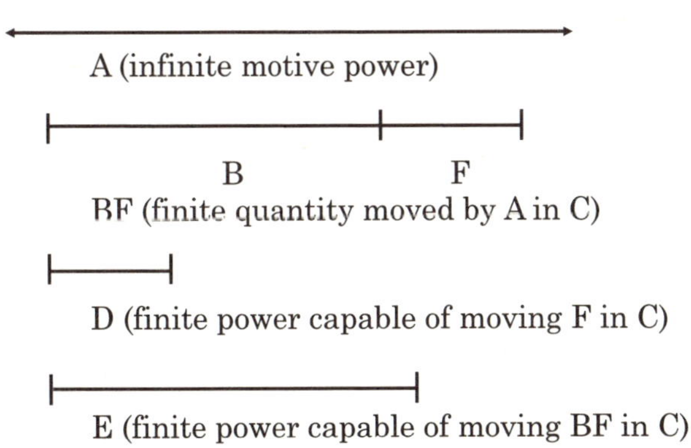

Fig. 2

But power D is able to move F in the time C; therefore E will also be
234,1 able to move BF in the time C. But the infinite A was also able to move BF in the time C. Therefore in the same time BF will be moved both by the infinite power A and by the finite E, which is impossible. Therefore the finite will not be affected in any way by the infinite. (He creates an unclarity in the passage by calling B – that which is moved by the infinite power – BF as the argument progresses, because it includes F, which was smaller than it, in itself).[238]

And concluding that it is necessary that if the finite is affected by the infinite then in the same time something infinite and something finite will both bring about the same alteration (which is impossible, because it is established that the greater acts in the lesser time), he infers that 'whatever time is assumed it will do the same', as if he had said 'even if you take a smaller or a greater time than C, it will do the same thing'.

Consequently there will be no finite time in which the infinite will cause motion, and so it will not cause motion at all, if indeed everything that causes motion must cause motion in time.[239] And it has been shown that it will not be moved either;[240] therefore there is no infinite natural body. But will the infinite yet move something finite in an infinite time?[241] But this is impossible too. For what undergoes alteration infinitely cannot be in a state of having been altered or moved, since it has no limit; and what does not admit of having come to be cannot come to be.[242]

> **275a24-b6** But no infinite can be affected by any infinite. [Let A and B be infinite, and the time in which B is affected by A, CD. But the part of the infinite which is E, since the whole of B is affected, will not be similarly affected in the same time: for let it be established that the smaller is moved in a smaller time. Let E have been moved by A in time D. Then E stands to some finite part of B in the same relation as D to CD. Thus it is necessary that this be moved by A in the time CD, since it is established that the greater and smaller things will be affected in a greater and a smaller time by the same thing, when they are determined in proportion to the time. Therefore there is no finite time in which it is possible for something infinite to be moved by something infinite – therefore the time must be infinite. But an infinite time has no end, while something that has been moved does have one. So if every perceptible body has either the capacity to act or be acted upon or both,] there can be no infinite perceptible body.

He next shows what remains [to be shown], namely that no infinite can be affected by an infinite, either in a finite or in an infinite time, and first that it cannot be so in a finite time. He proves this through impossibility too, making use of the axioms that the smaller is moved by the same power in a smaller time, and that in the case of unequal things moved by the same power the same proportion holds both between their magnitudes and the times in which they are moved.[243]

The proof is as follows. Let an infinite mover A be assumed, and an infinite B which is moved by A, and a finite time CD in which B is affected by A. Then take a part of B, namely E. Then since the whole of B is affected by A in the time CD, it is clear that the part of it E will be affected by A in a shorter time, since it is axiomatic that the smaller thing will be affected by the same thing in the smaller time. Then let there be a time D, a part of CD (in which the whole of B was assumed to be affected by A), in which E, a part of B, is affected by the infinite A.[244]

But then let us make E stand in the same relation to a part of the infinite [B] as the time D stands to CD, which is a finite part of something finite: for whatever ratio D stands in to CD, it will be possible to take the magnitude E as standing in the same ratio to another finite magnitude, which will itself be a part of B. Let this be F, although Aristotle does not assign this letter to it in the passage.[245]

As the time D is to CD, the magnitude E will also be to F, and, by alternation, as the time D is to E, CD will be to F.[246] But the time D stands in such a relation to E as being that in which E itself is moved by the infinite A; so, in the time CD, F will also be moved by the infinite A, but it was laid down that the whole of the infinite B was moved by A in the time CD, which is impossible, since the greater is moved by the same power in a longer time, and the infinite B is greater than the finite F.[247] Consequently this, namely that the greater and the smaller can be moved by the same thing in the same time, has also been reduced to an impossibility.

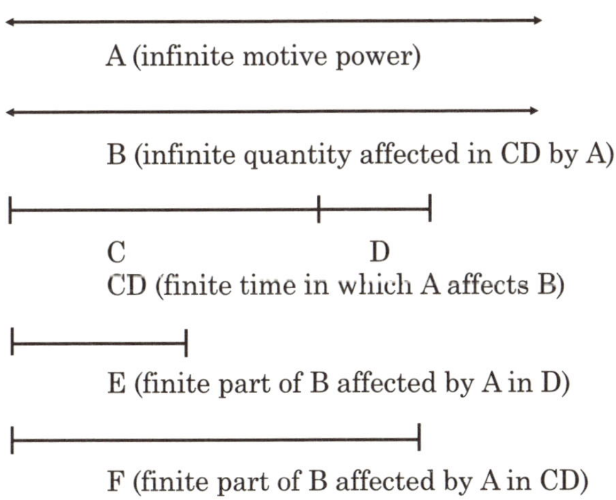

Fig. 3

But 'since the whole of B is affected, [it][248] will not be similarly affected in the same time', i.e. the greater and the smaller will not move the same amount in the same time, because it was established that the smaller was moved in a smaller time by the same thing, just as the larger and the smaller are affected by the same thing in a greater and a smaller time, in cases where the things which are affected and which act are determined in accordance with the same temporal proportion.[249]

Consequently neither the greater nor the smaller is simply affected without qualification, but rather things determined in accordance with temporal proportion are [affected] to a greater or lesser extent. So if it is impossible for the greater and the smaller to be moved by the same thing in the same finite time, he reasonably concludes that 'therefore there is no finite time in which it is possible for something infinite to be moved by something infinite'.

But could the infinite perhaps still be moved by the infinite in an infinite time? He briefly suggests that this is impossible by saying 'but an infinite time has no end, while something that has been moved does have one'. For something either being affected or acting in an infinite time, proceeding to infinity, does not advance towards an end,[250] and so cannot be said to have acted or to have been affected. And what is not of a nature to move to an end if nothing prevents is not capable either of beginning to act or be affected, as was said earlier;[251] for that which cannot have come to be cannot be coming to be.

Consequently he infers reasonably from what has been said the conclusion of the whole argument in the second figure, as follows: every perceptible body has the capacity to act, or to be affected, or
236,1 both; no infinite body has the capacity to act, or to be affected, or both; therefore no infinite body is perceptible, and, by conversion, no perceptible body is infinite.[252] ('Perceptible' was added to contrast with mathematical, since a mathematical infinite is frequently assumed.)[253]

> **275b6-11** But again, all bodies in a place [are perceptible. Therefore there cannot be an infinite body outside the heaven. But nor yet is there one up to a certain point. Therefore there is no body of any kind outside the heaven, for if it is intelligible, it will be in place, since 'outside' and 'inside' signify place. Consequently they will be perceptible.] Nothing is perceptible [which is not in a place].[254]

He showed earlier with many arguments that no body is infinite, and said that it follows upon this to investigate whether even if the total body is not infinite, none the less there is still enough of it for many heavens. But having set this on one side,[255] he once again shows by way of more general and more concrete[256] demonstrations that there is no infinite body. And clearly the argument concerns perceptible body, for it certainly [does not concern] mathematical [body].

He concluded that it was impossible for any infinite body to be perceptible; and if this is the case so too is the converse, that no perceptible body is infinite.[257] So, in addition to this universal negative premiss he now assumes an affirmative one which states that all spatially-located bodies are perceptible, from which he infers the conclusion that no spatially-located body is infinite.[258]

He assumes a further minor premiss to the effect that what is outside the heaven is spatially-located; and he concludes that there is no infinite body outside the heaven.[259] But he omits the conclusion of the first syllogism, that no spatially-located body is infinite, although positing as premisses first that no infinite body is perceptible and second that every spatially-located body is perceptible.[260] A little later he posits the minor [premiss] which states that what is outside the heaven is spatially-located, 'since "outside" and "inside" signify place', and so by way of this he will finally prove that there is no infinite body outside the heaven.

Alexander, however, draws the conclusions in the second figure[261] as follows: everything spatially located is a perceptible body; the infinite is not a perceptible body; therefore the infinite is not a spatially-located body. And then again:[262] the infinite body is not spatially located, what is outside the heaven is spatially located; therefore the infinite body is not outside the heaven. But it is clear that wherever it is possible to reduce arguments to both the first and the second figures, the first is to be preferred, if indeed the arguments in the second figure are completed by way of the first.[263]

And this analysis of the argument, which takes the conclusion and assumes in addition another premiss, concludes according to the third so-called '*thema*' of the Stoics, of which the structure, according to the 237,1 early [Stoics], is as follows: whenever from two [propositions] a third is deduced, and the conclusion along with some other independent [proposition] entails something else, the latter can be deduced from the original two plus the independent additional assumption.[264]

So from the [premiss] stating that every spatially-located body is perceptible, and that stating that no perceptible body is infinite, and from the assumption added to this conclusion stating that what is outside the heaven is spatially located, he concludes that there is no infinite body outside the heaven.[265]

But this had also been shown by way of the earlier arguments; for if there is no infinite body at all, clearly there will not be one outside the heaven. But as he is shortly going to show[266] that there is only one, unique world, and that there are no others besides it, whether finite or infinite in number, he prepares the way for this generally by showing that there is no body outside the heaven whether infinite or finite, and consequently that there are no other worlds besides this, whether finite or infinite in number.

So, having shown that there is no infinite body beyond the world, he then posits in addition that there is no finite one either (which he calls 'one up to a certain point': for a body which extends 'up to a certain point' and not in every direction endlessly is finite). He now posits this as undemonstrated, completing the disjunctive proof as follows: if there is a body outside the heaven, it is either infinite or finite; but it is neither infinite nor finite; therefore there isn't one. A little later on, he will prove this.[267] Thus having demonstrated one and assumed the other he draws the conclusion.

Therefore there is no body at all outside the heaven, be it intelligible or perceptible. For if someone were to say that there was an intelligible body outside the heaven,[268] he will locate it in space, since 'outside' and 'inside' signify place. But every spatially-located body is perceptible, as was already established.[269] For the bodies are located according to some perceptible difference, ultimately that in virtue of the inclination according to which each moves towards its proper place and remains [there].

And given that it is assumed to be outside the heaven, if it is encompassed by place, then it must have both tangible surface and apparent colour, and it will no longer be intelligible but perceptible. But there is no perceptible body outside the heaven, since there is no infinite [body], as was shown, nor any finite one, as was assumed.

Therefore two absurdities result from the assumption of an intelligible body outside the heaven. The first makes it no longer intelligible but perceptible (if indeed it is outside the heaven, which Aristotle too made clear by way of 'consequently they will be perceptible', and clearly no longer intelligible); but it follows from its being perceptible that it doesn't exist at all, if indeed it is neither infinite, as was proved, nor finite, as was assumed.

It is possible, Alexander says, that 'but nor yet is there one up to a certain point' (which is the same as saying 'not finite') is not stated simply as an assumption. For it has been shown[270] that every perceptible body is either that above, or that below, or that round the outside,
238,1 of which the upper and lower are within the heaven, while that round the outside is the heaven itself. So if there is no perceptible body besides these, there will not be any perceptible body outside the heaven.

275b12 It is possible to argue more formally as follows.

The 'more formal' arguments are so called by contrast with the concrete ones.[271] Concrete are those drawn from what is immediately relevant to the matter in hand, and which argue from the matter itself, so that it is not easy to apply the demonstration to something else. The formal ones, which they also call 'dialectical', are also true, but are more general,[272] and capable of applying to other cases, and are rather drawn from reputable opinions,[273] the method of which Aristotle supplied in the treatise on *Topics*[274] (he calls the commonplaces in respect of which arguments are constructed 'topics').[275] So having earlier shown, by way of proper and relevant arguments derived from the natural motion of each of them, that no simple body is infinite, he now proves the same thing by way of more general arguments.

> **275b12-18** No infinite body which is uniform can move in a circle, [since there is no centre to the infinite, and what moves in a circle does so around a centre. On the other hand, neither can the infinite move in a straight line, since it would require one equivalently infinite place towards which it could move naturally,] and another such towards which [it could move] unnaturally.

Having assumed that every natural body moves spatially, given that nature is a principle of motion, he shows that no infinite body, whether continuous or divided, is capable of moving spatially, from which he concludes that no body, whether continuous or divided, is capable of being infinite.

First in the case of continuous [bodies] he shows that the thing which moves in a circle, being uniform and simple, is not infinite, drawing his conclusion in the second figure as follows: the uniform thing which moves in a circle has a middle; the infinite does not have a middle.[276]

This more general proof is not derived from natural motion alone, and does not apply only to natural [bodies], or only to spherical bodies, or only to heavenly bodies, but to anything which moves in a circle, be it natural or artificial, spherical or of some other shape, heavenly or terrestrial. For everything which moves in a circle must move around a middle.

In the earlier arguments, he showed, on the basis of the motions which belong to them by nature, that neither heavenly nor sublunary bodies can be infinite. For neither could the revolving body move in a circle if indeed it were infinite,[277] nor could the sublunary [bodies] move in a straight line, since they would possess either infinite weight or lightness, which was shown to be impossible.[278] And the demonstration from the proportionality of magnitudes, motions, and times,[279] is appropriate in the case of naturally-moving bodies.

'Which is uniform' is added not because the non-uniform and composite [bodies] which revolve do not move around a middle (since everything which revolves moves around a middle), but because the argument here is concerned with simple bodies.[280] 'Which is uniform', in the sense of 'simple', is added, since he constructs the demonstration in this case, because in the case of composite [bodies] too natural motion takes place in respect of whichever of the simple bodies predominates in it.[281]

He shows that the infinite will not move in a straight line either, on the basis of the following *reductio ad impossibile*. If the infinite moves in a straight line, since things which move in a straight line move naturally whenever they are moving towards their proper place, and unnaturally whenever [they are moving] in the contrary direction, there must then be two infinite places, one to which it moves naturally and the other to which [it moves] unnaturally. But it is impossible for there to be two infinite places, since there can be neither two infinite bodies nor two infinite places, if indeed the infinite is everywhere.[282]

Thus he demonstrated that the infinite is double from the natural and unnatural motions of things which move in a straight line. And the same thing can be shown if the infinite moves in any way from one place to another; for there will then be two [infinite] places, one which it leaves and one which it moves towards.

This proof seems more general insofar as it applies as much to light things as it does to heavy. And he has already laid down this same argument, when he said 'but in fact it is not possible for the infinite to move in any way whatsoever, since it will either move naturally or by force; and if it moves by force, there will be some other [movement] natural for it, and consequently some other place equal in size to it towards which it will move – but this is impossible'.[283]

Alexander draws attention to this [passage], saying that it would be more reasonable for it to come here since it is more formal [in nature]; for what is at issue here is [the fact] that it cannot move in the manner of something which moves in a circle because it has no middle,[284] while there[285] he made no mention of this. But perhaps the proof can also be derived from what is proper to the things which move in a straight line (for proper to them are both natural and unnatural [motion]); and from the things which belong in common to both light and heavy things (for natural and unnatural [motion] belong in common to them). So he reasonably placed it both among the arguments drawn from specific characteristics, and among those drawn from general ones. The thing which moves in a circle is here treated as something general.[286]

> **275b18-25** Furthermore, whether it has its motion in a straight line by nature [or whether it moves by force, either way it will require an infinite motive force, since the force of an infinite body will be infinite, and an infinite force will require an infinite body. Consequently the mover will also be infinite (and there is

> an argument in *On Motion* that none of the finite things has an infinite power and that nothing infinite has a finite power). So if whatever can be moved naturally can also be moved unnaturally,] there will be two infinites, both that which moves in this way, and that which is moved.

Having shown that the body which moves in a straight line is not infinite from the [fact that] there would be two infinite places,[287] 240,1
which is impossible, he now proves the same thing from the similar impossibility of there being two infinite bodies. And he proves it on the basis of two propositions which he has already demonstrated: everything which moves is moved by something (which he demonstrated in Book 7 of the *Physics*);[288] and if the infinite moves, it is moved by an infinite power, but no finite magnitude possesses an infinite power, which he demonstrated in Book 8 of the same treatise.[289]

So the infinite moves in a straight line either naturally, or unnaturally and forcibly, and in both cases, being infinite, it will be moved by some infinite power of the body which moves it, which body must then itself be infinite. For infinite strength belongs to an infinite body, and the strength of an infinite body is infinite: this is now treated as axiomatic, since it was demonstrated earlier.[290]

And since the things which move naturally, even if they are moved by something, possess a mover within themselves which is incorporeal (although this [fact] contributes nothing to the derivation of the absurdity which shows that there will be two infinites),[291] he reasonably added 'so if whatever can be moved naturally can also be moved unnaturally', and then concluded 'there will be two infinites, both that which causes motion in this way, and that which is moved'. And it was also reasonable to add 'in this way';[292] for it is a consequence of something's being moved by force that there is something else apart from it which causes its motion.

So since what moves naturally can also be moved unnaturally, and something of this sort has an external mover, and if when what moves is infinite the mover must be infinite too, he reasonably said that, whether the infinite moves in a straight line by nature or by force, it will be moved by something infinite, not because it moves naturally [when it does so],[293] but because what moves naturally in a straight line can also be moved forcibly.

It follows from this that it is moved by some external infinite mover,[294] so that the upshot of the argument would be as follows: the infinite body, whether it is said to move in a straight line naturally or unnaturally, is capable of being moved forcibly and unnaturally; a body that is moved forcibly is moved by some body external to it which moves it. Something which is so moved externally, if it is infinite, will be moved by something of infinite power. But an infinitely powerful body is itself infinite in magnitude. Therefore the infinite body will be moved by some other infinite body. Therefore there will be two infinite bodies, a mover and a moved, which is impossible.

But while he had proved in the earlier arguments that no infinite could be affected in any way by an infinite, from relevant and appropriate [considerations] derived from proportionality, here he is happy to reduce the argument to something incredible (namely that there are two infinites), and for this reason he called the argument 'more formal'.[295]

241,1 **275b25-9** Again, what is it which moves the infinite? [If it moves itself it will be ensouled. But could this be possible, that there were an infinite animal? But if something else moves it, there will be two infinites, the mover and the moved,] differing both in form and capacity.

Having shown, Alexander says, the absurdity that follows from the infinite's moving forcibly (namely that there must be another infinite outside it which moves it), he now shows that even if it moves by nature the same absurdity follows.

For, he says, if everything which moves by nature is moved by something, either by itself or by something else, then if the infinite moves by nature it is moved either by itself or by something else. But if by itself, it will be ensouled, since the only self-movers are those which are moved by the souls in them. But if this is the case, the infinite body is an animal. But it is impossible for an animal to be infinite, since an animal is shaped and limited by shape, while the infinite is not shaped, and an animal has its parts symmetrically arranged with one another, while the infinite has no symmetry.

So if it is not moved by itself but by something else, there will be two infinites, the mover and the moved, differing from one another in shape, that is, separated by their own peripheries. For if they were continuous, they could not be affected by one another, since nothing can be affected by itself. And they will differ in power, if indeed one is a mover and the other is moved, both being infinite. Therefore there will be two infinites in reality, which was said earlier[296] to be impossible.

It is possible, Alexander says, both that the earlier remarks form several arguments and that they are all parts of a single argument, with him [sc. Aristotle] saying that since everything that moves is moved either by itself or by something else, and either naturally or unnaturally, and showing that something impossible follows in each case. For if [it is moved] unnaturally, then there will be several infinite magnitudes, since a mover of this sort is external and forcible. But if naturally, and if by itself, there will be an infinite animal, which is impossible, and if by something else, then there will once again be another infinite body, which by its own infinite power will move the infinite body.

It might appear, Alexander says, that he has passed over those things which move naturally, but do so neither because of a soul nor because they have something external which moves them: that is to say if they move because of their internal power. 'Or has he passed over this', he says, 'on the grounds that it has already been shown because of the impossibility of there being either infinite heaviness or infinite lightness?[297] Or because the argument is general (for he made the division on the basis of general opinion)? Or because he has shown in the *Physics*[298] that even these things are moved by some external mover (for it is the cause of generation for them)?' In this way, then, I think, and not very appropriately, Alexander [interprets] 'whether it 242,1
has its motion in a straight line by nature or whether it moves by force':[299] for he [sc. Aristotle] does not say that he is passing over [any of] the things which move by nature in this division.

So perhaps Aristotle employs the following division: the infinite body, if it possesses motion, has it either naturally or forcibly, and if naturally, either by nature in the manner of those things called natural, such as earth and fire, or in the manner of those things called self-movers, things moved by the souls within them. For 'forcibly' is distinguished by division from both of these.

So having shown that [it does] not [move] forcibly (for there must be some other external infinite mover), he shows that [it moves] neither in the manner of the natural things (because things which move naturally can also move unnaturally, that is to say once again forcibly, as a result of an external mover), nor in the way that self-moved things are moved by themselves, since it is not an animal. So it is necessary that [it be moved] by something else, and once again as a result of an external mover.[300]

275b29-276a17 But if the universe is not continuous, [but is divided[301] by the void as Democritus and Leucippus say it is, there would necessarily be only one motion for all things. For they are distinguished by their shapes, while they say that they have a single nature, as if each of them were a separate piece of gold. But for these things, as we said, it is necessary that there be the same motion, since wherever a single clod moves, there the whole earth goes, and all fire and a single spark both move to the same place. Consequently none of these bodies will be absolutely light, if all of them have weight (nor heavy if they all have lightness). Moreover, if they have weight or lightness, there will either be some extremity for everything, or some centre. But this is impossible, since it is infinite. In general, where there is neither centre nor extremity, neither up nor

> down, there will be no place for the movement of bodies. And if this does not exist, there will be no motion, since things must move either naturally or unnaturally, and these things are defined by the places proper to them, and those foreign to them. Furthermore, if there is a place where something rests or moves unnaturally, it must be natural for something else (this is confirmed by induction). So it is necessarily not the case that everything is heavy or everything light, but some are one and some the other.] So it is evident from these considerations that the body of the universe is not infinite.

Having shown that no natural, simple, continuous body, such as the elements are, can be infinite, he goes on to prove that neither can the elemental bodies be infinite in number as distinct [bodies], as Leucippus and Democritus[302] before him, and Epicurus[303] after him, supposed.

For they said that the principles[304] were infinite in number, and they thought them to be uncuttable, indivisible, and unaffectible on account of their being compact and having no share of void. For they said that division occurred because of the void in the bodies,[305] and that these atoms are separated from one another in the infinite void, and, differing is shape and magnitude and position and order,[306] they move through the void and meet with and collide with one another, and that some bounce off in whatever way they will, while others become intertwined with one another because of the congruity of their shapes, magnitudes, positions, and orders, and it so happens that the generation of composite things is accomplished.[307]

Arguing against such an infinity of elements, Aristotle draws a number of conclusions. First, if the atoms differ from one another in shape, magnitude, and such like, while their constitution and nature are the same (since their motion is according to nature[308] and not a result of their shape or magnitude), all of them must have the same motion, so that it will not be the case that some of them are heavy and others light, but all will be similarly heavy or light, which is evidently contrary to the appearances, and he will show later on that some natural bodies are heavy and others light.[309]

243,1 That they must have a single nature and a single inclination he shows also by way of examples, namely that of gold. For whatever the shapes of various gold nuggets, the inclination of all of them is the same. And [he shows it] on the basis of the simple bodies, 'since wherever a single clod moves, there the whole earth goes', even though there is such a difference in magnitude, and probably in shape as well. And 'all fire and a single spark move to the same place'. Consequently in the case of the atoms too, wherever one of them goes, all of them will go, and so all bodies will be either only light or only heavy. For if the simple [bodies] are like this, clearly the composite [bodies] will be as well: the composite [bodies] are in fact differentiated in inclination according to the predominance of the simple [bodies] within them.[310]

He next employs the following second argument based upon what has already been shown. If the atoms have weight, there will be a middle of everything towards which the heavy things move, and if [they have] lightness, there will be an extremity towards which the light things move. But it is impossible for the void, being infinite, to have either middle or extremity; therefore it is impossible for the atoms to have either weight or lightness. So if every natural body moves either because of weight or because of lightness, and the atoms have neither weight nor lightness, clearly they will not move.[311]

And he employs a third argument, also based upon what has already been shown, as follows. In the infinite void, there is indeterminacy. Wherever there is indeterminacy, there is neither middle nor extremity. Where there is neither middle nor extremity there is no up or down either. Where there is neither up nor down there is no place. Where there is no place, there is no spatial motion. Therefore there is no spatial motion in the void.[312]

And while the other things are plain, he proves that in the absence of place there is no motion as follows. If what moves moves either naturally or unnaturally, and both natural and unnatural motions are determined by their places (given that natural motion is motion towards the proper place, and unnatural [motion] towards somewhere foreign) – then if in the absence of place there is no determination in respect of place, there would be no moving thing either.[313]

Fourthly he infers 'furthermore, if there is a place where something rests or moves unnaturally, it must be natural for something else'. Alexander says that this is connected with the first argument:[314] for having shown there that it follows for those who hold that the atoms are of a single nature that all bodies must be either only heavy or only light, he now shows that this is impossible from the [fact that] a place which is unnatural for one thing is natural for another, which he confirms by induction.[315] For the place which is unnatural for earth, namely the upper one towards which it moves by force, is natural for 244,1
fire. And this being the case, if moving downward is unnatural for the atoms (since they are light), downward motion would be natural for some other body. But if upward [motion] were unnatural for them, upward [motion] would be natural for something else of another nature. And if this is the case, there must be some other bodily nature for which the place and motion unnatural for the atoms is natural. For thus does Alexander most ingeniously [interpret] his [sc. Aristotle's] conclusion that 'so it is necessarily not the case that everything is heavy or everything light, but some are one and some the other'.

But perhaps if Aristotle had argued this in connection with the first argument, he would not have prefixed the 'furthermore' to it after so long an interval; rather it appears that this [argument] too is indicative of the same thing, namely of the non-existence of motion in the absence of place, which is the principal subject of the passage.[316]

For if [the place] where something either remains at rest or moves towards unnaturally must be the natural place of something else, it is clear that there will be neither natural nor unnatural motion or rest in the absence of [natural] place. But in the principal argument he showed this on the grounds that natural and unnatural motions are determined by place, while here [he does so] on the grounds that what is unnatural for one is natural for another.

And the supposition stating that 'if there is a place where something rests' is uttered in relation to what is about to be said, not what has been said already. For [the argument] could not be completed by 'it must be natural for something else (this is confirmed by induction)', as though 'so it is necessarily not the case that everything is heavy or everything light, but some are one and some the other' is to be inferred from some other axiom; rather he would have said 'if there is a place where something rests or moves', etc.; 'it is necessarily not the case that everything is heavy or everything light'.[317]

And now, having shown on the basis of this argument too that if there is motion and rest, both natural and unnatural, there must be both a proper place and one foreign, he infers as a conclusion the remaining consequence of what has been said, 'so it is necessarily not the case that everything is heavy or everything light, but some are one and some the other'.

For if the natural and the unnatural are determined by their places, and if what is unnatural for one is natural for another, it is necessary that not everything be heavy and not everything light. For if there is a unique inclination, it would not be the case that some things are moved naturally and others unnaturally, since the place would be unique as well. So it would not be the case that what was unnatural for one would be natural for another.[318]

These arguments, then, utterly refute the view that the primary elements are atoms similar in nature and carried through an infinite void, as Leucippus and Democritus supposed. For if some motion is natural and some unnatural, and the places towards which the motions [occur] are determined, then movement will not [take place] in an infinite void, and the movers will be differentiated from one another in nature, and not only in shape and magnitude.[319] However the argument does not seem to have refuted the claim that, while not continuous, the body of the universe is infinite in mass. So how can he say that it is clear from this that the body of the universe is not infinite?

Aware of this objection, Alexander says that 'it can be shown that it is not possible for there to be an infinity of atoms by also employing what was said by him a little earlier on. For he said that "on the other hand, if they are dispersed, it may still be possible for the fire of all of them to be infinite".[320] And we will say the same thing about the atoms, so that if the single body composed of the atoms is dispersed in all directions, there will no longer be void. For if there is an infinity of triangular bodies, the infinity made up of them will not allow any room for bodies of other shapes. But it was possible, having already shown that the universe was not infinite in magnitude, to prove that it is not numerically infinite either, because what is composed of an infinity of things must itself be infinite.'[321]

However, Aristotle is clearly not here arguing against the idea that the atoms are infinite in number, nor are his conclusions drawn against this; rather some [are drawn] against the infinite void and others against the supposition that the atoms themselves are uniform in nature. So perhaps having first assumed in his concrete[322] demonstrations that a magnitude [composed] of magnitudes finite in number is itself finite, while that [composed] of infinites is infinite, and having shown that the universe is finite in magnitude, he proved that it is not [composed] of things infinite in number.

But here, having answered in more formal terms those who hold that the universe is both continuous and infinite, he turned to those who say that the universe is divided by void and composed of atoms of a uniform nature carried through a void; and answering them in more general terms he refutes the claim that it is composed of such things, whether they be infinite or finite in number. For it is clear that the conclusions drawn can be drawn even if the atoms are said to be infinite in number.

Reasonably, then, he did not invoke the supposition of infinity in saying 'but if the universe is not continuous, but is divided by the void as Democritus and Leucippus say it is', since both the infinite and the finite in number are encompassed by the term 'divided'. And in the conclusion he says 'so it is evident from these considerations that the 246,1
body of the universe is not infinite'. For it was demonstrated at the same time that the infinite could not exist as divided either, as Democritus and Leucippus said, given that it was shown that it could not be so divided at all, whether it be [composed] of things infinite or finite in number.[323]

[CHAPTER 8]

276a18-22 Let us now say why there cannot be more heavens. [For we said that this needed to be examined, in case someone might think that it has not been shown in general concerning bodies that it is impossible for any of them at all to exist outside this world, but that] the argument has only been made in respect of those with no definite position.

After having shown that there is no body of infinite size on the grounds that neither the heaven nor the sublunary elements are infinite, he sets out next to investigate whether even if the body of the universe is not infinite, it might perhaps yet be large enough to be capable [of generating] many worlds from itself.[324]

But having set out this problem, he resolved first of all to investigate in general if there can be an infinite natural body – not investigating those in the heavens or the sublunary [sphere], but more generally if there is any infinite body, as he does in the third book of the *Physics*.[325] And having shown that there is none, he demonstrates additionally that there is no body, infinite or finite, perceptible or intelligible, outside the heaven. And he here employs more formal approaches, showing again that there is no infinite body, neither a revolving body nor any of those which move rectilinearly.

And, after concluding that the body of the universe is not infinite, he next undertakes the postponed investigation into whether perhaps, even if neither this heaven nor the world dependent upon it, nor in general any body, is infinite, it might yet be large enough for many worlds. This is the same as the former question of whether this heaven or world is single and unique, or whether there can be many of them (by 'heavens' he now means those worlds which also have sublunary parts).[326]

It is clear that if someone were to think they had already demonstrated that in general it is impossible for there to be any body whatsoever outside this world, he will no longer investigate whether there is any other material world or not. But if someone thinks that 'the argument has only been made in respect of those with no definite position', he will likely investigate whether there are other worlds.

Alexander understands 'with no definite position' to refer to the infinites, and interprets as follows: 'in case anyone', he says, 'does not think that we have demonstrated in general that there is no body apart from the ones of this world, out of which it is composed,[327] which has indeed been shown, but only thinks that our argument is concerned with there being no infinite body. For by "with no definite position" he means the infinite, since the infinite has no determinate place, because, being infinite, it too is indefinite. We will show next that there is no body outside this world, from which it follows that there cannot be more worlds.'

But perhaps 'with no definite position' does not refer to the infinite, since he would not say this in the plural, for there are not many infinite [bodies]. Moreover, I do not think that 'in case someone might think that it has not been shown in general concerning bodies that it is impossible for any of them at all to exist outside this world' refers to those from which the world is constructed, since these[328] would not be contrasted with those 'with no definite position' as general to particular. Rather he is talking of that which he does proceed to show, namely that there is no body at all outside this world, from which it follows, as Alexander agrees, that there cannot be more worlds.

I think he talks of things with no definite position by contrast with those which contribute to the disposition of the world.[329] For if it is only shown that things with no definite position cannot exist outside the heaven, it is perhaps unclear whether those which contribute [cannot], and hence whether there are other worlds. So if someone does not think that this has been shown generally, now it should be shown that there is no body at all outside this world.

And it is reasonable of him to think that this now needs demonstration. For he has shown in what went before that no infinite body exists outside the heaven, given that it has been shown in general that there is no infinite body.[330] But by assuming, either as something agreed or as something that will now be shown, that there is no finite body (which he called 'one up to a certain point'),[331] outside the heaven either, he concluded[332] in general that 'therefore there is no body of any kind outside the heaven either'.

Alexander said earlier and now says again that it is possible to say non-hypothetically[333] 'but nor yet is there one up to a certain point', which is equivalent to 'nor is there a finite one', given that it has been shown that every perceptible body [moves] either up, down, or in a circle. And this was shown from the simple motions, of which the up and down are within the heaven while the circular is that of the heaven itself.[334]

> **276a22-7** All things, then, both rest and move [both naturally and forcibly. They move naturally to the place where they rest without force, and they rest [without force] in the place towards which they move [naturally]. They move forcibly to places in which they rest forcibly, and they rest forcibly in the places towards which they move forcibly.] Furthermore, if this motion is enforced, its contrary will be natural.

He now assumes as evident two axioms for what is to be shown. The first is that all natural bodies 'both rest and move both naturally and forcibly', and they move naturally to the place where they remain naturally, and conversely they remain naturally in the place to which they move naturally, and contrariwise they move forcibly to the place where they remain forcibly, while they remain forcibly in the place where they move forcibly.

The phrase 'everything, then, both rests and moves either forcibly 248,1
or naturally' can refer, says Alexander, to everything which is of such a nature as to rest, since those things which are of such a nature as to rest both rest and move both unnaturally and forcibly. Sublunary things are of such a kind; for that which moves in a circle can neither be moved nor be at rest unnaturally.[335] And from these axioms he

constructs general demonstrations concerning sublunary things, as we will see.

But if one is compelled to interpret 'everything which rests and moves' generally, it can mean, I think, that one sort of rest of things at rest and one sort of motion of things in motion is unnatural, while the other is natural; while the things which have both rest and motion have in each of them both natural and unnatural.

The second axiom is 'if this motion is enforced, its contrary will be natural'; he means that the contrary to some unnatural motion for something will be natural for it. Alexander supposes that there can be many unnatural motions for a single thing on the grounds that what is not natural is *ipso facto* unnatural; for a clod of earth is not only moved upwards unnaturally, but laterally as well. He thinks that by making this assumption he can rescue Aristotle's position: for he says that the many unnatural motions are not contraries,[336] but the contrary of an unnatural [motion] is *ipso facto* natural.

But he makes more difficulties for himself besides these. For if he says that there are many unnatural [motions], while to each of them there is a single contrary,[337] and the contrary of what is unnatural is natural, then there will be many natural motions for each simple body, which is absurd, since it has been shown that there is [only] one for each of them.[338] And if someone were to say that there was a single contrary to a plurality of unnatural [motions], no longer would there be one contrary for each thing, which is absurd as well.[339]

He resolves it by saying that it has been shown in the first works[340] that everything which does not move naturally moves unnaturally and contrarily, since everything that changes changes to its contrary: for even what is in between is in a way contrary.[341] Consequently the natural motion of something is the contrary of its unnatural motion.

It appears that Alexander says this for these reasons, namely because in the case of the simple natural motions, just as there is one natural [motion] for each thing, so too there is one contrary unnatural [motion] to each natural [motion], while all the others are mixed, and can neither be called unnatural nor natural without qualification. And if it is said that what is not natural is *ipso facto* unnatural, it is not said in the sense that what is unnatural is contrary to what is natural, but simply insofar as it is other than the natural.[342]

249,1 **276a27-30** So if the earth moves forcibly from there to the centre here, [then it will move naturally from here; and if what has come from there rests here without force, then its motion here will have been natural too.] For there is one natural motion.

Employing some earlier premisses, according to Alexander, he argues something along the following lines: 'for any earth which is outside this world, if it is carried towards the centre of this world forcibly and unnaturally, will move from here to there naturally; for it has been established that the contrary of the unnatural is natural,[343] and [motion] away from the centre of this world is the contrary of that towards the centre. And so upward [motion] would be natural for it, since upwards is away from the centre.

On the other hand, if the earth outside this world rests at the centre of this world naturally and not forcibly, then it will be carried to this centre naturally as well. For it has also been established that where something rests non-forcibly, it moves there [non-forcibly] too.[344]

And having laid these things down', he says, 'he adds "for there is one natural motion" by way of indicating that if it moves there naturally, it will not move naturally towards anything else. And having laid these things down', he says, 'he assumes in addition what is required, namely that if there are many worlds they will be [constructed] from bodies which are similar to one another in form.'

These are the very words Alexander wrote in explicating this passage. He seems to me to think that two absurdities are generated by the argument, first that if the earth out there moves naturally from here to there, upward motion[345] will be natural for it even though it is earth, and second that if it moves naturally in this direction[346] it will not move naturally in any other.

But perhaps Aristotle does not add 'there is one natural motion' for this reason: for the motion is single whether it be towards the centre of this world or towards that of another, as they are similar in form and constructed from bodies which possess the same capacities, as he goes on to say.[347] Nor, I think, does he reduce the argument to this absurdity, namely that upward motion will be natural for earth. For he infers this in the succeeding argument, when he says 'if this were the case, earth would have to move upwards in its own world, and fire towards the middle'.[348]

But if my conjecture is worth considering, having assumed what he goes on to show,[349] namely that if there are many worlds they will be similar to one another and composed from the same bodies, Aristotle reduces the argument to the absurdity that there will not be one motion for each of the simple bodies, but two. For if the earth from another world (being similar in form to that here and being possessed of the same inclination, as he will shortly show)[350] were carried 250,1
towards the centre of this world, it would be carried forcibly and unnaturally.[351]

And it is clear that it will be carried naturally towards the centre of the other world, since it is established that the contrary of unnatural motion is natural. So there will be two natural motions for that earth which are contrary to one another, that away from this centre and that towards that centre.[352] But if movement towards that place is not natural for the earth from there, but resting here is, then this [sc. motion towards here][353] is natural for it: for if motion from here to there is unnatural for something, then resting here is natural for it, and something is naturally carried to where it is naturally at rest; for this was already established.[354]

Then again, what is naturally carried here will have two natural motions, that away from the centre (i.e. there) and that towards the centre (i.e. here), which is absurd. But although he passes over this, he adduces the reason for the absurdity when he says: '*for* there is one natural motion', as the causal connective 'for' makes clear.[355]

276a30-b18 Furthermore, all the worlds must be made from the same bodies, [as they are similar in their nature. But then again, each of the bodies must have the same capacity; I mean fire and earth, and things intermediate between them. For if things in those worlds are so called homonymously in comparison with those in ours, and not because they have the same form, then the whole would only homonymously be called a world. So it is clear that one of them must move naturally away from the centre, while another moves naturally towards the centre, if all fire is the same in form as all other fire (and so for each of the others), just as the particles of fire are in this world. That this is necessarily the case is obvious from the hypotheses concerning the motions; for the motions are limited, and each of the elements is so called in relation to one of the motions. Consequently, since the motions are the same, the elements too must everywhere be the same. Therefore the particles of earth in other worlds too will be of a nature to move towards the centre of this one, and fire there will move towards the extremity of this one. But this is impossible, since if this were the case, earth would have to move upwards in its own world, and fire downwards,[356] and equally the earth would naturally move away from the centre here as it moves towards the centre there,] seeing that the worlds are situated thus in relation to one another.

Alexander says that these premisses are added to what has already been said in order to make another necessary supposition, namely that if the worlds are many they are [constructed] from bodies which are the same in form. But perhaps this is another argument, as the prefixed 'furthermore' makes clear, one which reduces to another absurdity the [idea of the] earth moving upwards and fire towards the centre,[357] since it makes them no longer earth and fire.

First he shows that anyone who says that there are many worlds must accept that they are [constructed] from bodies which are the same in form, differing only numerically, with earth, fire, and those [elements] in between them having the same capacities, fire in every case possessing heat and lightness, earth cold and heaviness.[358] He shows this as follows: those who say that there are many worlds say that they are similar to one another, so that they are the same in form, differing only numerically. So if they are alike in nature, they are [constructed] from bodies which are the same in form and which possess the same capacity, i.e. from fire and earth and those in between[359] and the revolving body. For what has been constructed from these things is considered to be a world.

But if someone says that the things here are merely homonymous with those in another world, and not synonymous and of the same form, necessarily the things constructed from them will no longer be the same in form, as was assumed, but will be called worlds only homonymously. So if they are supposed to be the same in form, what is constructed out of them must be the same in form as well. But if they are worlds only homonymously, we must explain each of their natures.

So, having shown this, he infers as a consequence of their being the same in form as those here that, given that there is fire and earth there too, the one will move away from the middle and the other towards the middle[360] in those worlds too, and that all fire is the same in form as all other fire, and so too with each of the other [elements], just as particles of fire are similar to one another in this world too, as are earth and the other [elements]. And thus from the fact that things in the other worlds are the same in form as those in this one, he shows that fire moves away from the middle and earth towards the middle in those worlds as well.

He proves the same thing also on the basis of the suppositions about motions which he laid down at the beginning of the book.[361] He called them suppositions in relation to what was shown on the basis of them, since it was there proved that necessarily all the simple movements there are are that away from the middle, that towards the middle, and that around the middle, given that there are two simple lines in accordance with which motion is generated, the rectilinear and the circular.[362]

Thus it was shown that the number of simple motions is determinate (for this is what 'limited' means), and that there are two rectilinear motions, and that each of the elements are given their form by reference to them. So if these are the only simple motions, not only here but also (if there are more) in the other worlds, and the elements are defined by their motions, it is clear that the elements and bodies would everywhere be the same.[363]

Alexander says that 'the motions are limited' might mean not only that the motions are finite in number, but also that each of them has a limit and is not generated to infinity: for what cannot have come to be does not come to be.[364] But it should be noted first that their not being generated to infinity has nothing to do with the present issue, and secondly that it is not true of revolution: for that proceeds to infinity, and is always in a state of having come to be as well as of coming to be.[365]

Having shown that the bodies in all the worlds will be the same in form as one another, he infers what follows of necessity from this even if there are many worlds, namely that the particles of earth in the other worlds would naturally be carried to the middle of this world as well, and those here to the middles of the others too, and similarly fire everywhere [would naturally be carried] to every extremity. For all bodies which are the same in form are carried towards the same place: for where there is one clod, there too is all the earth.[366] And both the middles and the extremities are the same in form as one another.

252,1 'It is clear', Alexander says, 'that earth everywhere and fire everywhere would naturally be carried here from their resting naturally in these places, if anyone puts them there.[367] For they will not indeed say that they do not remain when put in these places since both the bodies and places are the same in form.[368] But something moves naturally towards the place in which it rests naturally. So if it is a consequence for those who say that there are many worlds that earth in the other world moves naturally towards this middle and that fire in the other one moves naturally towards this extremity, and this is impossible, as he will show, then it is clear that that, of which this is a consequence, will also be impossible.[369]

For if the arrangement of the worlds is such that there is contact between them at the extremities,[370] if earth from another is to be carried towards the middle of this world, it must be carried upwards towards the periphery of that one in order to transfer into this world and be carried towards its middle, and fire there travelling from there and arriving in the outermost heaven of this world must descend from it to the sublunary region.

And for this to happen earth must be carried upwards in its own world and fire towards the middle and downwards in this world; and similarly earth, since it is carried from here to there, will be carried upwards naturally from the middle here towards the middle there, which is impossible: since it is earth [it cannot] be carried [naturally][371] away from the middle [here]. And these things turn out this way on account of the many worlds (if indeed there are [many]) being in contact with each other at the extremities and touching one another.'[372]

I understand the phrase 'since if this were the case, earth would have to move upwards in its own world, and fire downwards'[373] as meaning that the earth moves upwards in its own world and the fire downwards in this one.[374] But Alexander, taking 'in its own world' to refer to both of them, says that even fire will move towards the middle in its own world in order to move upwards in this one.

'For', he says, 'if some world were placed under this one, with this one on top of it, then since the fire directly opposite the contact

between the worlds would be beyond the earth there, in order to move towards that [place] which is upwards in this world it must first move with a [motion] diametrically opposed to its natural [motion] in the world where it is, at any rate if it is to move naturally in a straight line towards the upper [place] in this one.[375] But, as it moves with this [motion], necessarily it will also be approaching the middle of that world, which is downwards;[376] consequently it will first move naturally towards the lower [place]. For it will surely not be the case that the fire touching or adjacent to it will move naturally towards the upper [place] in this world, but that which is beyond it[377] will no longer 253,1
do so, since everything moves naturally in the same direction.'

Alexander is compelled to say this since he takes 'in its own world' to refer both to fire and earth, unnecessarily in my view. And I am surprised at his positing another spherical world under this one, with this one above it: for what will determine upwards and downwards for the worlds in relation to each other, unless perhaps we call the direction below our feet downwards and that above our heads upwards? But in the antipodes things are the other way around. And what is the point of the placing below and above? Even if they were side by side, the same things would result.[378]

But Alexander was right in his further assumption that in addition to their being moved with a motion that is unnatural and contrary to their natural [motion], the contrary motions, both upwards and downwards, will at the same time be natural for them, which is impossible. For it was laid down that, of contrary motions, if one is unnatural for a particular body, the other will be natural for it;[379] and furthermore there is a single natural motion for each of the simple [bodies].[380]

'For this reason', says Alexander, 'the bodies in this world will also migrate [to that one], if indeed the middles and extremities there are natural for the [bodies] here too. For the same places are proper for things which are the same in form.[381] So if, on the assumption of there being many worlds, it follows of necessity that contrary motions will at the same time be natural for the simple bodies, and this is impossible, then the hypothesis too is impossible.[382] For if someone were to say that earth from the other world moves towards the middle of this one forcibly and not naturally, he will say that motion away from this world is natural for it, and that is upwards and away from this middle.'

The upshot of the whole argument is as follows. If the worlds are many, alike in nature, and not [just] called worlds homonymously,[383] all of them must [be made] of bodies which are the same in form and differing only numerically. But if this is the case, the bodies will possess the same capacities and the same natural motions, so that fire and earth and the intermediates in those worlds will be just as they are here.

But if this is so, clearly some[384] will be of a nature to move away from the middle, the others towards it. For they are identical in form to those here, just as the particles of each of the ones here[385] are [identical in form] to one another. Moreover, there are only two simple rectilinear motions, that upwards and that downwards, and not just in this world, but wherever there is natural motion.

So if things are thus, and of the [bodies] which are the same in form in each world some move away from the centre and some towards it, but the middles and extremities in the worlds are the same in form
254,1 and not in name only, it is clear that the particles of earth in another world will be of a nature to move towards this middle too, and fire there towards this extremity.

And if this is so, earth must move upwards away from the middle naturally, and fire towards the middle and downwards. So if these things are impossible, the original hypothesis of which it is a consequence, namely that there are many worlds, will be impossible too.[386]

> **276b18-21** So either we must not assume that the simple bodies have the same nature [in the many heavens, or if we do say this, we must then make one middle and one extremity.] But if this is the case, there cannot be more than one world.

Having shown the absurdities which follow for those who say that there are many worlds constructed from simple bodies which are the same in nature, and which have the same motions as well as many middles and extremities, he now demonstrates that those who say that the simple bodies are the same in nature must also hold that there is one middle and one extremity, not only in form but also in number.

For the movement of heavy and light things will be towards numerically the same middle and extremity in order that the aforementioned absurdities do not follow from the postulation of numerically many middles and extremities, namely that earth will move upwards and fire downwards, and each of the simple bodies will have two contrary natural motions.

And if these things are absurd, something even more absurd will follow from them, namely that the simple [bodies] will no longer be what they are said to be, so that earth will no longer be earth, and fire no longer fire. For if the substance and essence of each of them consists in their having a particular natural motion, changing this necessarily changes their substance as well.[387] So if the middle and the extremity are one in number, there cannot be more than one world, since each world must have a middle and an extremity. Consequently for those who say that the same capacity applies to the simple bodies in all of the worlds it follows that they should say that there is one middle and one extremity: but for this there will be one world and not more than one.

> **276b21-6** But to suppose that the simple bodies have some other nature [if they are removed to a lesser or a greater extent from their proper places is irrational. For what difference will it make to say they are this far or that far removed? They will differ in proportion (the more so the further away they are),] but their form will be the same.

He resolves an objection that might be brought against what has been said, namely that earth in another world might not be carried naturally towards this middle on account of its great distance [from it], and if this is so, the absurdities which were inferred for those holding that there are many worlds will no longer follow, since earth will not be carried upwards and nor will fire be downwards. 255,1

But to ascribe the abandonment by the bodies of their proper capacities to the distances, he says, makes no sense: for no other nature (i.e. another natural motion) will result for the simple bodies by their being more or less removed from their proper places. For what difference does this or that degree of separation make in this world? They will differ only in that things moving towards their proper places from far away will start their movement more feebly, and in proportion: more feebly the greater the length of the distance.[388] But the form of the motion remains the same, whether the distance be greater or smaller.

And so in the same way those in other worlds, if indeed there are any, would begin their motion towards the places here more feebly in proportion to their distance, but would not exchange the form of their natural motion, which is their substance; for it is unreasonable to ascribe generation and destruction to the length of the separation.[389]

But nor will the fact that there are certain things in between which are dissimilar in form (e.g. the fire both there and here between the earth there and the middle here) be a reason for earth there to abandon its natural motion towards this middle. For not even in this world if one were to postulate fire as an intermediate between the earth occupying both the upper and the lower place, will that earth be prevented from moving naturally towards the middle.

Furthermore, as he himself previously said, if the earth from that world which were placed in the middle of this one would remain at rest – for why would it be disposed to move away? – clearly it will also move naturally towards the middle here, since it was established that something moves naturally to where it is naturally at rest.[390]

> **276b26-9** Furthermore, each of them must have some motion, [since it is obvious that they move. Shall we say that all their motions are enforced, even contrary ones? But something which is in general not of a nature to move] cannot move forcibly.

That there is a natural motion for each of the simple bodies has been shown, I think, right at the beginning,[391] where, having demonstrated that there are three natural motions on account of there being three simple lines,[392] the circular [motion] was shown to be appropriate for the fifth body,[393] and the two rectilinear [motions] for the sublunary [bodies], about which more will be shown in the third book of this treatise.[394]

But he proves it now since it is useful for him both in regard to what follows and what has already been said. For, as a result of this axiom, 256,1 the earth in the other world and the fire here will be compelled to move naturally towards this middle and this extremity.

Alexander says: 'if there were not some natural motion for the bodies, things might be moved by other things in all sorts of directions and form many worlds, since they are not moving towards the same places.' But perhaps things moving in a disorderly fashion would make disorders (*akosmiai*) rather than many orderly worlds (*kosmoi*).[395] So he shows that there must be some natural motion for the simple bodies (for this must be understood in the context), by assuming as evident that the simple bodies would move in all the worlds, and in the same way as[396] they are observed to move in this world since the worlds should be assumed to be similar to each other, and that natural bodies are most particularly known through their motion, while mathematical objects are unmoved.[397]

Thus he virtually poses the question: do the simple bodies in the worlds move, when they do move, naturally, or only forcibly and unnaturally? If they do so naturally, then we have what was sought, namely that there must be some simple motion for them. But if unnaturally, they would move in every contrary direction, such as away from the middle and towards the middle.

It is easy to show that this is impossible on the basis of what has already been established. For it was established that the contrary [movement] to unnatural and forcible movement is natural.[398] But he did not assert this, both because it was something which had already been said and because it was in need of argument, while he did not assert that what is not in general of such a nature as to move cannot move forcibly, because it was clearly self-evident.[399] For if it is not in general of such a nature as to move, it would not move at all, while if it moves in any way at all, it is clearly of such a nature as to move.[400]

So there is in general a natural motion for those things which are of such a nature as to move, with which they are of such a nature as to move. And the argument is a remarkable one: for everything unnatural derives from what is natural, and if the natural did not exist, the unnatural would not subsist.[401]

276b29-277a9 So if there is some natural motion [for them, given that they are the same in form and particulars, their

> motion must be towards one single place, such as for instance *this* middle and *this* extremity. But if [the places] are the same in form, but many [in number] (because as individuals they are many although each of them is indistinguishable in form [from the others]), none of the parts will be such in regard to one [place] and not to the others, but will be the same for all. The same is true for all things that are formally indistinguishable from one another, but which differ in number the one from the other. What I mean is this: if the particles here are similarly disposed both to each other and to those in another world, then one taken from here will not react differently towards those in another world from the way it does to those in its own, but similarly,] since they do not differ from one another in form.

Having shown that there is a natural motion for each of the simple [bodies], he shows that it follows of necessity from this that given that all of them 'are the same in form and particulars, their motion must be towards one single place, such as for instance *this* middle' in this world and its extremity, or towards those in another world.

For if heavy things naturally move towards the middle, either they will all move towards numerically the same middle, or [they will move] towards middles which are one in form, but numerically different. So if one were to say that, just as heavy things and light things in themselves are each one in form although numerically many, so too some move towards this extremity and others towards others which are formally identical but numerically different, and just as the movers are in relation to one another, so too there will be many worlds for the movers formally identical to one another, but numerically different and separate – so, he says, if one were to say this, one must also hold that the particular particles are not of one kind in one place and of another [kind] in another.

For a particular clod [of earth] will not be of one kind in virtue of its being here and another in virtue of its being in another cosmos, but the same. For all things which do not differ in form are similar to one another, albeit different numerically, and it is not merely the case that the things here [are similar] to one another and the things there likewise, but things here are also [similar] to things there. And a particle of earth or fire taken from here will exhibit no differences either in relation to particles which are the same in form here or in relation to those in another world.

So, just as the particles of earth here, which are the same in form and differing only in number, move towards this numerically single middle, so too things formally identical to them in another world, if indeed they differ from these [only] in number, will move towards the same middle. For if all the parts of earth are the same in form but different [only] numerically, why should these ones move naturally only towards this middle and towards no other, and others formally identical to them, differing only numerically from one another and from these ones, not move towards this, but towards something else the same in form as it? For if the numerical difference of the movers is the reason [for their moving towards] numerically different places, each of the heavy things hereabouts, which differ only numerically, would have to move individually towards different middles.

Alexander says that the places in the worlds are not determined prior to the difference of bodies, as stable entities which attract the moving bodies towards themselves, so that for this reason the bodies in each world move towards their own appropriate and distinct places. Rather the differences of place are determined by the natures and inclinations of the bodies, in accordance with which each of them moves naturally. For it was not so determined and ordered; rather downwards is simply this direction towards which earth and heavy things naturally move, and similarly upwards is that towards which light things [move]. Consequently if the difference in places is subsequent to the difference of the movers, everything which is the same in form from the outset will move towards numerically the same place, if nothing prevents it.

I am at a loss, in regard to Alexander's account, to see how, if the place is the limit of the encompassing body insofar as it encompasses what is encompassed,[402] the differences of the places are determined by the natures and inclinations of the enclosed bodies, rather than the other way around. For even if we say that downwards is the [direction] towards which earth [moves], and upward that towards which fire [moves], descriptions of this sort are merely indicative.[403]

The essence of the heavy is to be of a nature to move towards the middle of the world, of the light [to be of a nature to move] towards the extremity: and Aristotle defines them as such here. In general, the middle and the extremity would exist even if one considered the intermediate elements[404] which move towards them not to exist. How then could the differences of places be defined by the differences of the bodies which move towards them?

But this is once again what was sought; and the upshot of the argument is as follows. If there is natural motion for the simple bodies, the natural motion of each of them which are of the same form must be towards a numerically single place; but the antecedent is true: so too, therefore, is the consequent.

He proved the minor premiss, that there is natural motion, immediately;[405] while he proves the conditional on the grounds that the things here which are of the same form and differ only in number (i.e. the many pieces of earth) move towards a numerically single middle, and the many fires [move] towards a numerically single extremity. So if the things in the other worlds are also of the same form as those here, it is clear that those there will also move towards the same [place] as those here.

And if someone says that the motion of those which are enclosed by a revolving body towards a place in another world is the same in form [as the equivalent motion here],[406] in addition to making the assumption that there are many heavens and many revolving bodies (which he will show to be impossible), he clearly admits that, if they were not prevented, they[407] would move towards this middle and this extremity, and consequently that motion towards these places is natural for them. I do not know why Alexander thought the passage contained something more incongruous.

> **277a9-12** Consequently, one must either drop these hypotheses, [or there is one middle and one extremity. But if this is the case, then it is necessary that there be only one heaven and no more on the basis of the same indications] and the same necessities.

Alexander says that 'these hypotheses' are all those that have been stated before, both recently and earlier, both that there is a natural motion for each of the simple bodies,[408] and that those of the same form move towards numerically the same place;[409] for it follows from these assumptions that the middle and the extremity are numerically single. And if this is case, necessarily too the world will be single and not multiple (for by 'heaven' he now means the world).[410] For it is clear 259,1
that someone who accepts these hypotheses must also accept their consequences.

Alexander, in resolution of certain objections urged against this argument, says 'someone who further supposes that there are many worlds already constructed by something and having within themselves a distinct upwards and downwards, must not say that bodies which are the same in form in the different worlds move towards places which are the same in form yet different in number, the heavy downwards in each, the light likewise upwards. For to say that is to beg the question, since what is being investigated is whether there are many worlds.

But by beginning with the natural movement of bodies, by which and through which there is such an organisation and world, if the same things, moving according to their proper nature, can move towards certain other, numerically different, places and construct an organisation in accordance with their natural motions, whatever they are in this world, they already allow that there can be many worlds. But if it were impossible at the outset to entertain this notion, and the possibility of there being many worlds were destroyed, this, which follows from it, would be impossible.'

But, for all that Alexander says, the hypothesis that there are many worlds is not yet absurd. For someone who is going to accept the refutation of a hypothesis may appear to assume it; but to assume things which tend towards the refutation of a hypothesis as though they too agreed with it: this is to beg the question.

Thus in this case, while assuming that there are many worlds is not yet to beg the question, assuming on the other hand that the bodies in the worlds are the same in form in such a way that heavy ones move towards the middle in each, and light ones to the extremity in each, and not towards a numerically single [place], is to beg the question. For the question is whether things the same in form move towards a numerically single [place], or towards different ones, so that the number of worlds can be established from this, and someone who assumes that they move towards different [places] begs the question, since it follows necessarily from this that there are many worlds.

Alexander says further, in resolution of another objection, that 'it is not the same to make such an assumption in the case of animals. For they think that, just as in the case of those which are numerically separate but the same in form, like men for instance, the parts of each of them too are identical in form to those of others who are numerically separate, the same will hold good in the case of the worlds.

But the assumptions are not the same. For while animals do not move in virtue of the nature of the bodies from which they are made (since there is another natural motion for each of them, according to which they move as animate creatures), none of the bodies within them move in virtue of their natures. Nor is the original composition of animals as a result of the local motion of their[411] parts.

For it is not as a result of the heart's moving here, while the liver, bones and all the rest of them [move] somewhere else, that the body comes to be formed, but each of them is in the place where it originally came to be; and nor do the things which nourish them move towards them in virtue of their own natural movements either. For neither the nourishment which is brought in from outside nor that which has already been prepared,[412] moves naturally towards them, but is prepared and carried to the parts by the nutritive soul.[413]

But bones, nerves, and flesh do not come to be in various places and then move towards their proper places which are identical in form to them. And nor, if a part of them is cut off and separated from those to which it is in contact, will it move back towards them once it is removed. For animals are not what they are in virtue of the natural movement towards their proper places of the bodies of which they are made; but the essence of a cosmos is the natural motion of the simple bodies.'

277a12-23 That there is a certain place towards which it is natural for earth to move, [and also fire, is clear from other

> considerations also. For, in general, whatever changes changes from something to something else, and these things, that from which and that to which, differ in form, and all change is limited: for instance, someone being healed [changes] from sickness to health, and something growing from smallness to largeness. Therefore so does the thing that moves, since it comes to be there from here. Therefore that from which and that to which it naturally moves must differ in form. Just as in the case of someone who is being healed, the mover [moves him] neither to any chance state, nor to where it [sc. the mover] wishes. Therefore fire and earth do not move to infinity, but to opposing places. But the upper is opposed in respect of place to the lower,] consequently these will be the limits of movement.

Having shown that the world is unique on the basis of the fact that there is a single middle and a single extremity (the middle towards which all the heavy things move and the extremity towards which the light ones [move]), he now proposes to establish this too, namely that there is a certain place towards which earth moves naturally, and one for fire. For if this is not established, it will be shown neither that there is a middle and an extremity, nor that middle and extremity are unique, nor that the world is unique in virtue of this reasoning.

He proves this by showing that the down and the up are limits of their motion, which in turn he shows by means of showing that their motions are limited and do not proceed to infinity, and this because all change is limited, because it is from something determinate to something determinate, and this because it is from opposite to opposite, and this because it is from one thing to another, from one thing to another differing in form.

Such is the analysis of the argument: it has the following synthesis.[414] Things which move in respect of place move from somewhere to 261,1
somewhere, since if they remain in the same place they do not change. That the change is from and to opposites differing in form he showed in the first book of the *Physics*.[415]

Moreover, motion from the intermediates occurs as if from opposite to opposite.[416] And proceeding sequentially, he first assumes that what moves moves from something differing in form to something [else] differing [in form], since no change could occur from one form to the same [form]; for the changing thing departs from the thing from which it changes.

And if [it changes] from opposite to opposite, [then it changes] from determinate to determinate, since opposites are contraries which are most distinct in the same genus,[417] and 'most' is determined. But if these things are determined, clearly they are limits, and that motion and change which occurs towards them will be limited too. However, with 'all change is limited' he appears to recall the demonstration[418] which says that what cannot approach something cannot move towards it in any way, and that what is moving cannot arrive somewhere infinitely distant from it.

And he establishes inductively that change occurs towards limited things by showing that in the case of things changing by alteration, and [things changing] by growth and diminution, both that from which and that towards which are different in form, determinate, and limited. Furthermore, natural growth and diminution (and generally change in respect of quantity) are determined, and everything else apart from them is unnatural.

And just as in the case of things changing in respect of quality and quantity the change is from somewhere to somewhere, and the state from which and the state to which differ in form and are contraries, determinate and limits, so too in the case of things changing in respect of place the differences of the place of things moving naturally – the up and the down – are different in form, contraries, determinate, and limited.

Therefore the natural motions of earth and fire will be towards these places. Therefore there is somewhere towards which fire moves naturally, and [similarly for] earth: the upper and the lower [places]. And 'the mover [moves him] neither to any chance state, nor to where it wishes', since someone who in general is being healed from illness, given that he is being healed, and even if the healer does not wish it, is changing in the direction of health: otherwise he would not be being healed.

> **277a23-6** Even motion in a circle has in a sense opposites, [the [endpoints] on its diameters; but for the motion as a whole there is no contrary. Consequently even for these things, motion is in a way] towards opposites and limited things.

262,1 Saying that rectilinear change in respect of place occurs from opposite to opposite (for this is most properly movement), he adds by way of superabundance that circular motion too (even though it is not movement in an unqualified sense but a rotation), since it is still a motion and a change, he says, 'has in a sense opposites, the [endpoints] on its diameters'. For this is the largest determinate distance in a circle, 'consequently even for these things, motion is in a way towards opposites and limited things'.

And if someone is puzzled by how Aristotle could both have demonstrated earlier[419] in many ways that there is no contrary movement for circular movement (and in particular that diametrically opposed things are opposed not as moving in a circle but are rather opposed as moving along the diameter), and yet could now turn about and say that [motion] in a circle 'has in a sense opposites, the [endpoints] on its diameters', first let him appreciate Aristotle's precision. For he does not say simply that [motion] in a circle has opposites, but that it 'has *in a sense* opposites', and that 'even for these things motion is *in a way* towards opposites'. For those things are most properly called opposed to one another which move towards contrary places, and the contraries are the ones which are furthest apart from each other.

But things which move in a circle do not move towards the place diametrically opposite as if to an end, but from the same place and to the same place;[420] and nor is motion along the diameter the greatest motion for the thing moving in a circle, for the greatest is that which admits of no addition, but after [completing] the distance across the diameter, the motion and the distance still admit of addition until returning to the same place.[421]

And, saying with great precision that 'even motion in a circle has in a sense opposites, the [endpoints] on its diameters', he adds 'but for the motion as a whole there is no contrary'. For there is the apparent character of opposition in the parts of the circular motion but not in the whole, and even in the parts not properly, if indeed properly-opposed motions are distinguished from rest and whenever one and the same natural mover moves in opposite directions, it moves one way naturally and the other unnaturally, while things that move in a circle do not come to rest when they reach the diameter.[422] For they do not move [along the semi-circles] in virtue of themselves;[423] and nor do they move naturally along one semi-circle and unnaturally along the other.[424]

277a26-7 Therefore it must have some end and not move to infinity.

Having shown that motions of place, and most particularly rectilinear ones (for the argument required them for the demonstration that there is one middle and one extremity) do not proceed to infinity, but terminate in opposite, determinate, and limited [places], and having added that circular motion 'has in a sense opposites', he draws the remaining conclusion, namely that the moving things 'must have some end and not move to infinity'; and if this is so, then there is somewhere towards which both earth and fire are of a nature to move, and if this is so, then there is one middle and one extremity.

And circular motion is unwearying and goes on to infinity, but not to the same kind of infinity as would something which moved in a straight line and always proceeded from one place into a new one; for circular motion is always in its end just as it is always in its beginning.[425]

277a27-33 An indication that earth cannot move to infinity [is that the closer it gets to the middle the faster it moves, while fire

> does so the closer it gets to the top. If it were infinite, its speed would be infinite too, and if its speed were, so too would be its heaviness or its lightness. For just as a body that is lower than another because of its speed is fast because of its weight, so if its increase were infinite,] the increase in speed would be too.

He adduces as an indication that the things which move in a straight line move to a determinate lower and upper [place] and not to infinity the fact that earth moves more quickly the closer it gets to the middle, as does fire the further up it gets.

For if this is the case, and the distance were infinite and the end not determinate, it would increase in speed infinitely. So if speed occurs because of the natural inclinations, greater heaviness in the case of the things which move downwards, greater lightness for those which move upwards, it is clear that both heaviness and lightness, being internal natural capacities, will also admit of increase to infinity, and if that is the case, then these moving things themselves need not, as Alexander thought, be infinite; but they must at any rate be receptive of increase to infinity.[426]

For since body is limited, and not receptive of increase for ever, it will be impossible for its proper capacity to be increased infinitely. So if earth and fire evidently cannot be increased infinitely, given that the recognized parts of earth maintain their same measures, clearly their speed will not increase either. And if the speed does not, then neither will their heaviness and lightness, since if the latter increase, their speed, as a concomitant of them, must increase too. And if these things are not the case, the progress of bodies that move in a straight line will not be to infinity, but to some determinate end.[427]

With great precision he calls the confirmation of the natural motions derived from the speed 'an indication', because the natural inclinations are the causes of speed, and confirmation derived from the effects is thought to be indicative.[428]

That there will be an increase of speed to infinity through the infinite increase in heaviness and lightness, if [this increase took place] in the case of one and the same thing, he shows by developing the argument in two ways. For just as something which is moving below something else, i.e. which takes the lead and moves downwards more quickly because of its greater speed is faster by the addition of heaviness, similarly too in the case of the very same thing, if the increment in heaviness is infinite, so too will be the increment in speed. The demonstration no longer derives from an indication but from a cause:[429] for the greater heaviness is the cause of the speed.

Whenever Alexander says that 'the moving things themselves would be infinite if their speeds and heavinesses and lightnesses were infinite', he should not be understood as meaning [that they are] actually infinite (since that is not a consequence of their movement towards infinity), but rather that they are increasing towards infinity.[430]

Having taken the argument to this point, that both the heaviness and lightness would increase to infinity, Aristotle was satisfied, on account of having previously shown that neither heaviness nor lightness can be infinite, that no finite body can have a capacity for infinite increase. Aristotle refrains from pointing out for us, perhaps thinking it to be well known, that if indeed heavy things move faster as they approach the bottom, and light things [move faster as they approach] the top, it is clear that these are their natural places, and that they move in an opposite manner towards their opposite [places].[431]

But while there seems to be general agreement in these matters that bodies move faster when closer to their proper places, different people give different reasons for it. Aristotle thinks that as they draw near they gain power from their proper wholeness, and take on a more complete form.[432] So by addition of heaviness the earth moves more quickly as it approaches the middle.

Hipparchus, in his work entitled *On Things Which Move Downwards Because of Their Heaviness*,[433] says that, in the case of earth thrown upwards, it is the throwing force which is the cause of its upward motion as long as it overcomes the power of the thing thrown, and it moves upwards faster in proportion to the extent to which it does overcome it. But when it diminishes, it first no longer moves upwards as quickly, and then moves downwards employing its own proper inclination even though some of the upward power still 265,1
remains along with it; and the more it fades, the faster the descending object always moves downwards, fastest of all when that power finally gives out.[434]

He gives the same explanation in the case of things dropped from above; for they too retain for a time the power of what held them back, which by counteracting [their motion] becomes the cause of the initial slower motion of the thing dropped.

'These things', says Alexander, 'might be well said in the case of things either forcibly in motion or forcibly at rest in an unnatural place; but it is no longer correct to say this in the case of things which [just] after having come to be move towards their proper place in virtue of their own proper nature.'[435]

In the case of weight, Hipparchus says the opposite to Aristotle: for he says that things are heavier the further they are removed.[436] But this too seems implausible to Alexander. 'For', he says, 'it would be much more reasonable for things changing from the contrary nature and becoming heavy instead of light still to retain something of their former nature at the beginning of their descent, while they are still coming to be and changing towards that form towards which they change in descending, becoming always heavier as they progress, but still retaining the power by which they were held aloft at the beginning and which impeded their descent.

The more so', he says, 'if it is natural for heavy things to be at the bottom, since it is for this reason that they move there. And they will be completely heavy and in possession of their proper form in this regard whenever they are at the bottom. And as they come nearer and approach the complete possession of their inclination, it is reasonable that they take on some addition and increment in regard to weight.

And it would be absurd, given that things which move and travel downwards invariably move more quickly naturally the further they are away from the top, to say that they do so when they are lighter; for this would destroy [the claim] that they move with a downward motion as a result of heaviness. If things did not occur in this way in all cases, it might be possible to say something. But as it is, the same thing is seen to occur in the case of things thrown upwards forcibly, and [in the case of] those kept aloft, and [in the case of] those which are changing in the upper [place] into one of the things for which downward motion is natural.'[437]

This is what Alexander says against Hipparchus; and he is particularly right in my view when he says that if speed is a result of heaviness, just as the motion of balances clearly takes place because the heavier [side] drags it down first,[438] so it would be impossible for it to be heavier when further away, and yet move more quickly and with greater speed when closer.

However, quite a few people say that the cause of things moving downwards faster the closer they get to the bottom is that the higher they are, the greater is the bulk of the underlying air which they ride on, and the lower they get, the less [the bulk of the underlying air which they ride on], and for this reason the heavier things move more quickly because it is easier for them to cut through the underlying air.

For just as things descending in water seem to be lighter as their downward motion is impeded and resisted by the water, it is reasonable that things behave the same way in air, and that the more underlying air there is, the more impeded they are, and hence seem to be lighter. And in a similar fashion a greater quantity of fire moves upwards more quickly, as it cuts through the overlying air more easily; and the greater the bulk of the overlying air the more sluggishly will [the fire] which is passing through it move.

For even if air does not have the same constitution as water, even so, being corporeal, it impedes the motion of things moving through it. And if this is true, the increment in speed will not occur, as Aristotle says it does, by way of increase in heaviness, but by removal of the impediment.[439]

But Alexander rightly remarked that, even if things were so, Aristotle's argument showing that the bodies have determinate places and do not move to infinity on the grounds of their moving more quickly the closer they get to their proper places remains unrefuted. For something travelling an infinite distance never gets any closer, and nor does the underlying air get any less (or more), since it is infinite, if indeed the distance is infinite, so that even if someone accepts that this[440] is the explanation of the swifter descent, he will even so show as a result of it that things cannot move to infinity.[441]

And if, provided that the distance is infinite, something descends more quickly not because of the decrease of the underlying air,[442] it is clear that it will move more quickly because of the increase in heaviness: for what other cause could there be for things moving naturally and not forcibly, as Hipparchus said?[443]

'A better', says Alexander, 'and more physical explanation for the faster movement of things closer to their proper places is that of Aristotle, which says that heaviness and lightness increase. And Aristotle says that this happens since something approaching its proper place is always more properly in possession of its form, and for this reason it becomes heavier (if it is heavy) and lighter (if it is light).'

I think one should first of all investigate what is said to be agreed by everyone, namely how it is that we know that near their proper
places things move more rapidly. For if this increase in heaviness or 267,1
lightness takes place, something weighed in air (if someone stretches out from a high tower, branch, or sheer cliff and weighs it in the air) should appear to be heavier when someone stands and weighs it on the firm foundation of the earth underneath; but this does not seem to be the case, unless one were to say that the difference in these cases was imperceptible.[444]

> **277a33-b9** But nor could they be moved upwards and downwards by something else, [nor forcibly (as some have it, 'by extrusion'). For then more fire would move more slowly upwards, and more earth more slowly downwards. But the contrary invariably occurs: more fire and more earth travel more quickly to their own places. Nor would it move more quickly near the end, if it did so forcibly and by extrusion. Everything gets slower the further it goes from that which is impelling it, and if it moves from some place forcibly, then it moves towards it without force.] Thus we can be sufficiently confident about what has been said after considering these things.

In demonstrating that the world is unique, he relied on the uniqueness of its middle and extremity, and he showed this from [the fact that] natural motions of bodies take place towards something determinate and limited towards which they are of a nature to move, and [he showed] this from [the fact that] bodies approaching their proper places move more quickly, he now shows that these motions of the bodies occur naturally, something he has made use of on many occasions, both in the recent discussions and in earlier ones.

He shows it by refuting the opposing theses. For the elementary bodies move either as a result of their own nature, or are moved by something else, or are squeezed out by one another. So he shows that they are neither squeezed out by one another nor moved by something else; first that they are not moved by something else, on the grounds that a greater bulk of fire moves faster than a lesser, and a larger clod of earth than a smaller. But if they were not of a nature to move thus and were moved forcibly by something, the greater bulk would be moved more slowly by the same thing, since the lesser is more easily forced and carried.

And if one were to say that the greater moves more quickly because it is moved by a much greater power, the argument would be fantastical. For it is not necessary that the greater always be moved by something greater,[445] but it always moves more quickly. And this is also clear from the opposite motion: for the greater bulk of fire is dragged down more slowly.

And he shows that they do not move under the force of mutual extrusion either as follows. This opinion was held after him by both Strato[446] and Epicurus,[447] who thought that every body possessed heaviness and moved towards the middle,[448] and that the lighter ones settled out above the heavier ones by being forcibly squeezed out upwards by them, so that if the earth were removed, water would move to the centre, and if the water [were removed] the air, and if the air [were removed] the fire.

Aristotle refutes this account also on the basis of the fact that extruded things which are further removed from the extruder that forces them invariably move more slowly in proportion to their distance away, while in the case of these bodies' motions exactly the opposite is seen to occur: for their movement invariably becomes quicker as they move away from the origin, and they move quickest of all when in the vicinity of the things towards which they are moving.

And one could use the aforementioned argument concerning things moved by something else in the case of extruded things as well. For just as the greater bulk will be moved more slowly by the same thing, so too will it be more feebly extruded: but more fire evidently moves more quickly upwards. But while he perhaps rejected this argument on the grounds of overkill, at the same time he has perhaps forestalled the objection which states that a larger thing is always squeezed out by something larger and possessed of greater power surrounding it, which is not necessarily the case for something which moves [something else].[449]

But having made separate replies to those who say that things are moved by something else and those [who say that] they are moved by extrusion, he adds a general argument against those who say that motion is forcible and not natural. For if things evidently move without force to that place from which they are moved forcibly, by whatever kind of force, it is clear that they move naturally. For it would not have been by force that they moved from that place in the first place if it was not natural for them to be there;[450] and something also moves naturally to the place where it remains naturally.[451]

Consequently for someone who says that bodies move by force and not naturally, it follows that he must also say that they move naturally. And if all motion were forcible for them, they ought to move equally in opposite directions; but now earth evidently moves upwards forcibly, but downwards by its own nature, and fire contrariwise. So he plausibly adduces in addition to these demonstrations the fact that it is possible for people theorising on this basis to arrive at sufficient conviction concerning what has been said, namely that there is a natural motion for bodies, which takes place towards their proper places and to what is akin to them.[452]

However, those who treat as an indication that everything moves naturally towards the middle the fact that when earth is removed water moves downwards, and when the water [is removed] the air [does so too], do not know that the reciprocal motion[453] is the cause of this. For when the denser things are transferred into the place of the rarer, the rarer take the place of the denser, propelled downwards because there can be no void, and because body cannot pass through body.[454]

But one must realise that it was not just Strato and Epicurus who held that all bodies were heavy and moved naturally downwards, unnaturally upwards, but Plato too knows that this opinion is held,[455] and disputes it, thinking that 'downwards' and 'upwards' are not properly applied to the world, and refusing to accept that things are called heavy in virtue of their downward motion.

He writes in the *Timaeus* as follows: 'it is quite wrong to think that it [sc. the world] is divided into two separate places which are completely opposed, the lower towards which everything which has bodily mass moves, and the upper towards which everything moves against its will.'[456] Moreover those who describe the atoms as dense say that they have heaviness, and are the cause of heaviness in composite [bodies], while [the causc] of lightness is the void.

But one must consider in all of these cases what Aristotle means by the lower place, if indeed place is the limit of the encompassing body, insofar as it encompasses what is encompassed.[457] Let the upper place be the interior of the sphere of the moon: but this is impossible, because there should be nothing above the upper place, while there are many things above the lunar sphere.

On the other hand, if the lower place is the surface of the earth, in the first place it docs not itself encompass anything but is encompassed; and secondly how can either earth itself or all heavy things move below the surface of the earth? For there is no place lower than the lower place. And if the lower place is the centre of the universe,[458] it is clear that the upper will be the periphery, and place will no longer be the limit of the encompassing body, insofar as it encompasses what is encompassed.[459]

But, it seems, Aristotle was the first really to attempt to investigate what place is, and he got hold only of a partial conception of it, as something which is external [to its object] in virtue of its encompassing [it], and for this reason made his account vulnerable to many objections.[460]

> **277b9-12** Moreover, it might also be shown by way of the arguments of first philosophy, [and those concerning the circular motion which is necessarily everlasting both here] and in other worlds as well.

He calls the treatise *Metaphysics* 'The Arguments of First Philosophy', and says that it can be shown from what was said there and in Book 8 of the *Physics*[461] that the world is unique. For in Book 8 he
270,1 shows that circular motion is eternal wherever it may be, either in this world or in some other, and that is caused by an immobile cause,[462] which was shown to be incorporeal and partless, on the grounds that no body could be infinite, while if it were finite it could not possess infinite power; but it is an infinite power which moves infinitely.

Alexander says that he shows in the *Metaphysics*[463] that this mover of the revolving body is single: for since it is immaterial and incorporeal, it will consist solely of definition and form which are numerically the same.[464] But if it is single, what is moved by it will be single as well, and, given that there is one revolving [body], necessarily there is only one universe.

Alexander puzzles in this context as to why one prime mover could not move more revolving bodies, at any rate if they move out of desire for it as an object of love;[465] for nothing prevents many things from desiring the same thing. And speaking as follows, he says 'or does this argument proceed in accordance with probability, and not as showing something necessary?'

But I think one must first note that Aristotle does not say that the mover of the revolving body is single, but rather that while the simple movement of the fixed [sphere] is caused by the first substance, each of the wandering spheres is moved by an unmoved, eternal substance.[466] He writes in Book 12 of the *Metaphysics* as follows: 'the origin and first of things is unmoved, both essentially and incidentally, and it causes the primary and eternal motion. Since what is moved must be moved by something, and the prime mover is essentially unmoved, and eternal motion must be caused by something eternal, and one motion by one thing, and since we see that besides the simple movement of the universe, which we hold to be caused by the primary unmoved substance, there are other eternal movements, the ones of the planets (for the revolving body is eternal and unresting), each of these motions must also be caused by an essentially unmoved and eternal substance.'[467] Furthermore, how could it be other than absurd for these arguments, which have been called in evidence from first philosophy and his most precise physical treatise, not to possess the necessity of demonstrations? But even if there are many moving causes, and there is one prior to all of them, necessarily there will be one organisation of everything in relation to a single cause of things organised, and there will be a single intelligible world from which and in relation to which the perceptible world is single and itself generated from containers and things contained, just as the intelligible is single.[468]

And if those who say that the many worlds are collected into one organisation, so that there will be one thing [made up] of all of them, some more in the nature of wholes, others more in the nature of parts, some as containers, others as things contained, what they say is probable, but it amounts to saying that a single world is composed of many worlds for each of the heavens and the things which fill the sublunary [region] is a world: thus earth is a world, as is water, air, and aether.[469]

So for this reason this world will be single, because the intelligible world which moves and creates it is single, having an ordered multiplicity, but a unified and causal multiplicity, just as Aristotle demonstrated the [multiplicity] of the primary, unmoved, intelligible, moving causes in the *Metaphysics*,[470] towards which he now directs the argument. For if the eternal revolutions are primarily caused by a single intelligible world constructed from many unmoved causes, it is clear that the same things will also complete the perceptible world as a single thing.[471]

Thus the argument derived from first philosophy and the eternal revolution is not merely probable, as Alexander thinks, and deprived of necessity; and nor does Aristotle merely suppose God to be the final cause of the world, which caused Alexander's problems, but the efficient cause as well.[472]

And one pronouncement of many of his in this book suffices [to show this], namely that God and nature do nothing in vain;[473] and it further suffices to show that the eternal motion is transmitted from there to the revolving body, which has of itself [only] a finite power. And it is also sufficient that our master Ammonius shows this throughout his book, namely that Aristotle understood his God to be not only the final but also the efficient cause of the world.[474]

And if someone who says it is only an object of desire is puzzled as to why many heavens cannot be moved by one object of desire, it is easy to say that multiplicity does not get close to the One, unless it is itself unified. For like is close to like. For this reason none of the individual parts is capable of being essentially unified by the universality of the One, but it is necessary that the things which proceed from the One first of all be united in respect of other things with all speed and with them to be extended towards the One.[475]

> **277b12-27** And it also becomes clear [to investigators in this way that it is necessary that there be only one heaven. Since, given that there are three bodily elements, there will be three places for the elements too, one for the sinking body around the middle, another for the body that moves in a circle, which is the extremity, and a third intermediate between them for the middle body. For it is necessary that what rises be there, since if it is not there it will be outside – but it is impossible for it to be outside. For some things are weightless, while others have weight; and the place of the body which has weight will be lower down, given that the place of the heavy is around the middle. Nor yet could it be there unnaturally, for then there will be something else for which this place is natural for it; but there is no other body. Therefore it must be in the intermediate place. Of this itself, what its differences are, we will speak later on. So concerning the bodily elements, what kinds there are and how many of them, and what is the place of each of them, and generally how many places there are,] is clear for us from what has been said.

Having undertaken to show that the world is single from [the fact that] for each of the elements there is some determinate proper natural place towards which they move and that this [place] is in this world, and having shown that there is a determinate proper place which is natural for each [of the elements] towards which its motion is determinate and limited, he confirmed this on the basis of the fact that moving things move more quickly near their proper places.[476] Then, having brought in evidence from first philosophy, where the number of counteracting[477] spheres also needed to be counted in order to find the number of unmoved causes to move them, he supplies what is missing from the argument, namely that the natural place of each of the elements is one in number.

So, having opened [this passage] with the problem to which the things proved appertain (which was that the world must be single), he goes on to show that the natural place of each of the elements is one in number, as follows.

There must be as many differences of place in the world as there are differences in respect of their locomotion [exhibited by] the bodily elements from which the world is constructed. So, as there are three bodily elements (at least speaking generically),[478] there will be three places as well, one around the middle for the heaviest body which settles under everything, another at the extremity for the revolving body which encompasses everything, and a third for that in between which is divided into more [parts]: into the light which rises above everything else which moves in a straight line, namely fire, and into the two bodies between it and the heaviest,[479] namely air and water.[480]

That these places exist in this world and no other he shows, I believe, in the single case of fire, as he has previously shown in the case of earth. For it is necessary that this rising thing, i.e. fire, either be in this intermediate region or be outside the place in which it lies.

This is how Alexander interpreted 'outside', not as being in another world, but in this one. 'For the same account', he says, 'would hold even if it were outside this world, since if it is there it will be somewhere and in place, and if in place, then in one of these three. As it is outside [the world] it will not be where the thing which moves in a circle is, since this place belongs to it, and it cannot depart from it. He shows that it cannot be naturally below [on the grounds that] that is the proper place of the body which possesses heaviness. And this', Alexander says, 'holds equally in this world and in another; for since there are three places in that one too, the same account holds in that world as well. For if someone says that the intermediate place there is natural for it, it is clear too that, since there is in this [world] an intermediate place which is the same in species as that place, it will also be naturally in this intermediate place in this world.'

Thus Alexander expounds it. However, it would not be to the present point to show that fire has this as a natural place in this world. It is better, then, if possible, to interpret 'it is impossible for it to be outside' as Aristotle's way of showing that it is impossible, for 273,1
there to be a place outside this world for the rising body.[481]

For if it were outside, it would be either higher than this world or lower. Since there are generically two bodies, the weightless, i.e. the light, and that which possesses heaviness, if the rising [body] were outside this world it would also be lower,[482] which[483] is the place, the lower place, of the body possessing heaviness (if indeed its proper [place] is at the middle), and not that of the rising [body]. On the other hand, how could the rising [body] here, if it were higher, get lower, since it is weightless, given that the lower place is that of the body possessing weight?

And if one were to say that the rising [body] is unnaturally either in this place where it now is (as Alexander interprets it),[484] or outside this world (as I've just finished arguing), either the place of this fire, where it now is said to be unnaturally, or its supposed place outside this world will be natural for some other body, since what is unnatural for one thing is natural for another.[485] But there cannot be another body besides these, the heavy and the light, as was shown from the necessity of the motions.

Consequently this intermediate place[486] is natural for everything light, if nothing light can be outside this world either unnaturally or naturally. From which it follows that the natural place of each thing is numerically one, and from this that the world is unique, which does not follow from the [fact that] the place it occupies is proper for fire, unless one assumes in addition, as Alexander does, that everything light will move there, which was what was at issue.

And saying that the rising [body] is in the place intermediate between the middle and the extremity (since 'intermediate' taken thus indefinitely will not only be the place of the rising [body], but also of air and water),[487] he reasonably adds 'of this itself', i.e. the intermediate, 'what its differences are, we will speak later on'; for in the final books[488] he discusses the [bodies] in this intermediate region and their arrangement.

Alexander takes this passage as indicating that Aristotle calls the etherial body an element too when he says 'given that there are three bodily elements': 'he says this', he says, 'either because it has the account of an element in relation to the composition and substance of
274,1 the world, or because all of the simple bodies are generally called elements, and because he wants the elements to be the simplest [bodies].'[489]

Immediately afterwards he concludes by calling the simple bodies elements, [which he does] even more clearly at the beginning of the third book.[490] He means by 'bodily' elements things which are bodies but are elemental, distinguishing them from form and matter: for the latter are elements of bodies, indeed genuinely and primarily so, but they are not elements in the sense of being bodies.

[CHAPTER 9]

277b27-278a16 We will show not only that the world is unique, [but also that there cannot come to be any more of them, and further that it is eternal, since it is ungenerated and indestructible; but first we should run through some problems concerning this, since it would seem to someone considering it in a certain way that it is impossible for it to be single and unique. For in everything which is constructed and generated either by nature or by skill, the shape in and of itself is different from the shape combined with the matter. For example, the form of a sphere is different from a golden or a brazen sphere; and equally the shape of a circle is different from a brazen or a wooden circle. For when we say what it is for a sphere or a circle to be, we do not include gold or bronze in the account, since these things are not part of the essence (although if it is a bronze sphere or a golden

> one, we will include it); and this is true even if we cannot conceptualize or grasp anything other than the particular. For sometimes nothing stops this from occurring, as for instance if only one circle were grasped; for it will no less be the case that the essence of circle and this particular circle, its form and the form in the matter as one of the particulars, are different. So since the heaven is perceptible, it will be one of the particulars: for everything perceptible exists in matter. But if it is one of the particulars, being this heaven will be distinct from being a heaven without qualification. Therefore this heaven and heaven without qualification are distinct, the one being as form and shape, the other as being mixed with matter. But for anything of which there is a shape and form] there either are or can come to be many particulars.

It is often the case that, while it is self-evident that something is as it is, it is by no means evident that it cannot be otherwise (so that if it does not now exist, it could never come to be). For it is easy to see that man is naturally five-fingered, but that it is impossible for him naturally to be otherwise is by no means understood by all.

So, having shown that the world is single and not multiple, he adds in the scientific manner[491] that it could not even be otherwise, so that not only is it [actually] single, it could never come to be multiple. For perhaps someone might hold that, while the world is now single, there is nothing to prevent it from becoming multiple. And, being about to show by way of other arguments besides the former ones that this world is eternal because it is ungenerated and indestructible,[492] he first of all rehearses a persuasive argument which apparently establishes that it is impossible for the world to be single and unique, or rather that it is impossible for it necessarily to be one and the same. For the difficulty entails that either there are many individual worlds or that many could come to be.

For this reason Alexander thought it right to understand 'necessarily' in the phrase 'it would seem … that it is impossible for it to be [necessarily] single and unique'. The difficulty is the following. In all things, he says, both those which exist by nature and those constructed by art, and in general everything [composed] of matter and form, the form considered in itself is one thing, while the composite of matter and form, which is both said to be and is exhibited in common in every individual thing (if there are more than one of them), is something else.

And even if there are not many individuals, but one only, such as 275,1
the single bird [called] the phoenix or the single circle, it is no less true that the form itself is one thing and the composite [of matter and form] another. For the circle is different from the essence of a circle,[493] which is considered in accordance with the circle's form, and different

again is this particular circle which exists in matter. In the cases where it is possible to separate conceptually shape and form from matter, in those cases the form is of a common nature, and there either are or can be more enmattered perceptual things which participate in it.

Reasonably, then, having assumed that the circle is one in number, he has shown that even in this case there is the same difference between the form which is conceptually separable and the composite [of matter and form]. For even if the world is unique, since it is enmattered and perceptible, the world's essence and its form, which is of a nature to be conceptually separable, is different from the composite, and for this reason the same one could be instantiated in many cases and there could be many worlds, just as, even if the circle were unique, there could for the same reason be many of them.

Further, in cases where there are many perceptible [instances], the commonality of the single form in all cases is clear. And even if the perceptible [instance] were unique, and for this reason we cannot conceive of something common and distinct from the individual (as if only a single circle were grasped), even then it would be no less true that the form, which is conceptually separable from matter and not dependent upon it, would manifest a common nature which was capable of being instantiated in many individuals which come to be because of matter.[494]

> **278a16-23** For if there are Forms, as some say, [this necessarily follows; and it does so no less even if nothing of this sort can exist separately. For, when we consider everything of this kind, that is such as to have an essence in matter, things similar in form are numerous, limitless indeed. Consequently either there are many heavens, or there could be. So from these considerations someone might infer both that there are and that there can be many heavens. We need to consider once again] which of these things has been well said and which has not.

He strengthens the difficulty to do with the commonality of form, since some (such as the Pythagoreans, Socrates, and Plato) say that the common is separated and transcendent,[495] thinking that there must be something prior to the many similar things which is the cause of their similarity and which transcends them, while others say that it is inseparable from the many things.[496] But whichever way you go, he says, in every case we see the same form existing in many enmattered things which are identical in form.

For many humans participate in the human form, and many horses [in the equine form], and so on for each of the others. And even if there are not many of them, still there could be because of the commonality of form, so that in the case of the heavens – or, what amounts to the same thing, worlds – there either are, or at least could be, many of them. For, says Alexander, the world considered without matter will be a genus or a species,[497] and genera and species are invariably predicated of many things.

But perhaps someone might say against this: what then? Is the form of Socrates, in respect of which his essence belongs to Socrates, 276,1 and which is conceived without the matter, a genus or a species? Or is Socrates' form not conceptually separable, with the composite Socrates rather existing as a single individual, and does not subsist in the same way as 'human' does?[498] For this reason Aristotle did not put down [in the passage] names of particular individuals, such as the sphere of Saturn or the circle of the Zodiac, but rather sphere and circle, which are applicable alike to gold, bronze, and the heavenly bodies.

And I am of the opinion that he now recalls the Ideas[499] because he wishes to strengthen the difficulty by way of division.[500] For whether forms are separable, as those who speak of Ideas [have it], or whether they are inseparable, many things clearly participate in each form. Some friends of Plato say he is alluding to Plato, since there ought to be many worlds insofar as they [derive] from the exemplars, given that we see many images generated from each exemplar. But Plato thinks he has demonstrated that the world is unique from the uniqueness of the exemplar.[501]

Alexander was also of this opinion, and chided Plato for having said that the world was unique because its exemplar was unique. 'For of everything else', he says, 'of which there are Ideas according to him, the Idea and exemplar is unique: for there are not at any rate many men-in-themselves.[502] Yet none the less there are many things that have their being in relation to it. For perhaps he rather shows from this the opposite of what he wanted. For if it were the case that, just as each of the intelligible animals[503] is numerically one so too were the perceptible ones, what was generated in relation to the exemplar which encompasses in itself all the intelligible animals would be single, and it would encompass in itself all perceptible animals similar to itself. But if there are an unlimited number of perceptible animals in accordance with each of the intelligible animals, it will no longer be necessary for what was generated in relation to the exemplar encompassing all the intelligible animals to encompass all the perceptible ones. For everything that was generated in relation to it would have to be one in form, but not indeed one in number. For just as there are many perceptible animals in accordance with each intelligible one, so too it will follow that there are many perceptible worlds if there is one exemplar.'

I am amazed how Alexander, in saying these things, could not understand that not all perceptible animals are such that there are many of them in accordance with each intelligible animal. For there are not many suns or many moons, but all the heavenly animals are unique.[504] So it should be said that each of the forms established in the world, not only those in the heavens but the sublunary ones too, of a single exemplar there is a single one.

277,1 For not only are the sun and the moon single, but the man which exists for ever in the world as an image[505] is single too, and so too is the horse, which have their being in things which are generated and destroyed; and these things are common and inseparable forms of matter, which Aristotle also accepted as being eternal, although assuredly not as prototypes.[506]

For none of enmattered [forms] is primary, given that matter is shaped by participating in the forms, and that if something participates in something, the participation must be one thing, and the thing from which the participation derives another. For this reason the forms in matter are not prototypes, but they are imprinted in the ultimate things[507] in virtue of their similarity to the prototypes.

Each of the generated and destructible things has something common, for instance being a man, [which derives] from the single exemplar, and being something particular, [which derives] from the disposition of the heavenly and the sublunary bodies changing this way and that, from which individuals come to be (since [the disposition] is also of an individual even if it is productive of many things) and to which it corresponds. And for this reason the wise delineate both the genera and the species and the shapes of generated things by looking to the configuration of the heavenly bodies, tracing out the similarities which the things which derive from there bear to them.[508]

Consequently each of the eternal things, as well as each of those which are generable and destructible, being one in number, corresponds to a unique immediate exemplar; and Plato well demonstrated that the world was unique on the basis of the uniqueness of the complete intelligible animal that was the world's exemplar.[509]

Alexander adduces further arguments to the effect that it is not merely possible, but necessary that there are many worlds, for which he does not supply the refutations, unless indeed he thought that the refutation supplied by Aristotle would resolve these ones too. Thus he says that the world was generated either by a creator or by nature. If by a creator, then it is reasonable that he could produce others too (as we see in the case of craftsmen); but if by nature, one can say the same thing. But it should be said that if the maker produces many, they must be either similar in all ways, or dissimilar in some ways. But if they are similar in all ways, a multitude of them would be redundant, whereas if they are in some way dissimilar, only one out of all of them would be perfect.[510]

He offers a second argument: if in the case of the other things which are imperfect and share only a little in being we see that the form perfects many of them, this would be all the more likely in the case of the best and perfect thing.[511] But each of the many things could not be perfect in all respects, given that there were other things besides it.[512]

He offers a third argument: if it is better for the best things to be
many, and the world is the best thing, and produced by the best, it is
reasonable for there to be many worlds. But it is not possible for many 278,1
things to be the best unless they are comprehended in one, because
the good and the best of all is single.[513]

> **278a23-b11** That the account [without the matter and that of the shape in matter are distinct has been well said; and let it stand as being true. But none the less there is no necessity on account of this that there be many worlds, nor is it possible for many to come to be, given that this one is made of all the matter there is, as it is. Perhaps what has been said will be clearer thus. Given that aquilinity is a curvature of the nose or of flesh, and the flesh is the matter of the aquilinity, if one flesh was made out of the all the flesh there is, and it were to be aquiline, there would be no other [instance of the] aquiline, nor could one come to be. Similarly, given that fleshes and bones are the matter for human beings, if some human being were to come to be out of all the available flesh and bone, and they could not be broken up, then it would be impossible for another human being to come to be. The same holds good also in the other cases; for in general it is not possible for any of the things whose essence is in some underlying matter to come to be if no matter exists. The heaven is indeed both a particular and a thing made of matter; but if it is composed not of some part of it, but of all there is, even given that there is a distinction between being a heaven and [being] this particular heaven it will not be the case that there either is another, or that many others could come to be, since it comprehends all the matter there is.] Therefore it remains to show this, namely that it is composed of all the natural and perceptible body that there is. But let us state first of all what we mean by '*ouranos*', and in how many senses, so that the object of our inquiry will become clearer to us.[514]

Here he begins to resolve the difficulty. The difficulty assumed two premisses, one holding that in the case of the heaven its form was one thing, the composite another, because it is perceptible and enmattered and one of the individuals, the other holding that in cases where shape and form are one thing and the composite another, of these either there are or there can be several individual instances.

He accepts the minor premiss,[515] but rejects the major as not necessarily holding universally. For it is not necessary, in cases where form is one thing and composite another, for them to be more [than one] in number. For if one thing were composed of all the matter there is, for instance if the form of the aquiline or the snub[516] nose in flesh were to come to be singly out of all flesh, and similarly for fleshes and bones (the matter of man), if a man were to come to be out of all flesh and all bones [there are], it would still be true in these cases that the form was one thing, the composite another, but it would no longer be necessary for the things participating in this form to be many in number, if the composite were of all the matter. Consequently it is not necessary that there be many aquilines or many men, since the major premiss does not necessarily hold universally.

With great precision, he adds 'and they could not be broken up': for even if a man were put together out of all the flesh and all the bones [there are], but these things could then be broken up, there would be nothing to prevent them being put together again to generate another man on another occasion, and for numerically many men to be generated.[517] But if they remain unbroken it would not be possible for there to be another man, given that man is enmattered, and there is no other matter besides that forming the basis for the one man.

Consequently too even though the heaven is enmattered and per-
279,1 ceptible, and for this reason one of the individuals, there is something else in it besides the composite, namely the essence of the heaven, and [the essence] not of heaven simply, but of *this* heaven, i.e. of an individual;[518] for the form in it is one thing, the composite another. But there will not be another universe on account of this, nor could many be generated, because this one has comprehended all the matter [that there is].

When Aristotle says 'for in general it is not possible for any of the things which are essences in some underlying matter', Alexander says: 'this should indicate that by "the form in the underlying matter" he means more generally that which requires a certain substrate; for this is his accustomed usage. Or does it stand', he says, 'for those things whose essence involves matter?[519] These things are clear, and this is reasonably said. But', he says, 'given that the form is in the underlying matter, the soul will also be in a substrate. So how can he say in the *Categories*[520] that no substance is in a substrate?'[521]

He resolves this well: 'for it is the composite', he says, 'the enmattered substance which is in and of itself, which is there said not to be in a substrate.'[522] In my view he brought up the example of the soul to no purpose, for the sake only of showing that even according to Aristotle the soul, considered as a complete actuality and form, is in a subject.

But in *On the Soul*[523] Aristotle says that there are two types of complete actuality and form, one separable the other inseparable, and in regard to the intellect and the contemplative faculty he says clearly that 'this seems to be a different type of soul, and this alone can be separated, as the eternal from the destructible'.[524] And in the third book,[525] talking of the active intellect of the soul, he says that 'this intellect is separable, unaffectible, and unmixed, being in essence actuality'.

Here, however, the argument clearly concerns the corporeal forms, of which he says that it is not possible 'for any of the things which are substances in some underlying matter' to be generated without there being some matter. Having said that the difficulty will be resolved through the fact that the world is composed out of all of the matter [there is], he reasonably proposes to show next that the heaven (or rather the world) is composed out of all of the matter [there is], by distinguishing first how many senses 'heaven' has, and saying that in one of its senses it means what we call the world, of which it is now undertaken to show that it is composed out of all of the matter there is.

278b11-24 In one sense, then, we call [the substance of the outermost orbit of the universe, or the natural body that occupies the outermost orbit of the universe, 'heaven'. For we are accustomed to call the outermost and highest in which we say that everything divine is located 'heaven'. But in another fashion we call the body which is continuous with the outermost orbit of all ['heaven'], where are the sun and the moon and some of the stars: for we say that these too are in the heavens. Yet again, we call the body which is encompassed by the outermost orbit 'heaven'; for we are accustomed to call the whole and the universe 'heaven'. So the word 'heaven' is used in three ways; and the whole which is contained by the outermost orbit must be composed of all the natural and perceptible body that there is, because there neither is] nor can come to be any body outside the heaven. 280,1

He says that 'heaven' has three senses. For we also call the sphere of the fixed [stars] by the special name 'heaven', which he defines in two ways, calling it the actual substance of the outermost orbit of the universe, i.e. the furthest revolving substance, or the natural body at the outermost orbit of the universe, taking the natural body in place of the more general substance, as being the more approximate genus of thing under discussion.[526]

And he cites in evidence the custom of calling the highest and outermost [part] 'heaven'. For if we say that everything divine is located in heaven and worship it as being located in the highest [place] of all (indeed he said at the beginning that everyone assigns the upper place to the divine),[527] clearly we think it appropriate to call the outermost and highest [place] 'heaven', even if we use the name for other [places] as well.

Alexander notes that he does not say 'outermost and highest' as if the upper place and the circuitous or the outermost place were the same, but as it is customary to call that outermost thing the highest, since he himself, opposing upwards and downwards, and saying that the light moves upwards while the heavy [moves] downwards, defines 'up' as being the concave [surface] of the lunar sphere: for that is where fire, the most rising and lightest thing of all, moves to.[528] 'The outermost thing', he says, 'would also be up, given that it is above the lightest [body].'

But how can we say that the divine is located in the [sphere] of the fixed [stars], if we think the wandering [spheres][529] are divine as well? Is it because we assign the very highest of all [places] to the divine, but treat the whole heaven as divine and designate it according to its highest part, and our turning towards the divine which goes towards the upper [part][530] in fact extends to the very highest [place]? Or is it because all things divine and transcendent depend upon the primary principle, and by their own nature are shared in by the primary and most beautiful of bodies?[531]

'Heaven' has a second sense, in accordance with which we call the wandering [body] 'heaven' as well. For this is the body continuous with, or rather next to the outermost orbit of the universe, in which [are found] the moon, the sun, and the other stars which are said to wander.[532] That we also call this 'heaven' is confirmed by our saying that these stars too are in heaven. In applying the appellation separately to both the fixed and to the wandering [bodies], he makes it understood that we also call everything which revolves and is eternal 'heaven', by contrast with what is generated and destructible. For he himself demonstrated in a general fashion in regard to it that, since the whole possessed one nature and one revolving motion, there was besides the sublunary elements another fifth substance of the heavenly body, possessing neither weight not lightness;[533] and he showed in general (and will later demonstrate)[534] that this was ungenerated and indestructible. And so he too evidently calls the whole too by this one name.

He says that the third sense of 'heaven' is that 'of the body which is encompassed by the outermost orbit', along with the outermost orbit itself as well, assuredly. 'For we are accustomed to call the whole and the universe heaven'; consequently the encompassed body is included along with the encompasser, and in this way the whole universe is called 'heaven'.

The [usage of the] word is customary with Aristotle's predecessors as well: Plato indeed calls the world 'heaven' in many places, for instance in the *Statesman* where he says 'that which we named heaven and the world,'[535] and in the *Timaeus*: 'indeed the whole heaven – or the world, or something else, for let us call it by whatever name it most properly receives.'[536] The world is also called 'heaven', either because it was fitting to name it after the most distinguished of the things in it, in the same way as man is called 'man' after his soul, indeed his rational soul, on account of his looking up at what he sees,[537] and his gathering many things into one: for alone among the animals man by reasoning gathers what comes from many perceptions into a single thing.[538]

And if one considers the etymology of the heavenly body which Plato offered in the *Cratylus*, namely that the heaven is 'the looking upwards,'[539] because having turned towards its proper causes and remaining there, the universe is plausibly called 'heaven' since it depends upon its causes.[540]

'So the word "heaven" is used in three ways'; and the third meaning, by which the whole universe is called 'heaven', is that in which, he says, it is said to be composed of every natural and perceptible body. And he adds the reason: 'because there neither is nor can come to be any body outside the heaven'. For if there is no body outside it, it is clear that it must be made of everything corporeal [that there is].[541] So this, then, remains to be shown: that there neither is nor can come to be any body outside the heaven.

278b25-279a2 For if there is some natural body outside the outermost orbit, [it must either be one of the simple or one of the composite bodies, and it must be there either naturally or unnaturally. But it couldn't be one of the simple bodies, since it has been shown that the body which moves in a circle cannot give up its own place; and nor yet could it be one that moves from the centre, or a sinking body. For they could not be there by nature (since other places are proper for them), and if they are there unnaturally, the place outside will be natural for some other body (since what is unnatural for this must be natural for something else). But there is no other body besides these. Therefore it is not possible for any of the simple bodies to be outside the heaven. But if none of the simple ones [can be], then none of the mixed ones [can be] either,] since wherever there is a mixed body, there must be simple bodies too.

If there is some body outside the outermost heaven, it must be either simple or composite. The simple bodies, as was shown from the simple motions, are the one which moves in a circle, and the rectilinear ones, the light one which rises from the middle and the heavy one which sinks towards the middle.[542] But the one which moves in a circle could not be outside this world because it was shown (says Alexander) that it could not move with any unnatural motion either,[543] and if this is so, then clearly it could neither be nor come to be in any place other than the circle where it is.

But it is clear that when he speaks of a 'natural body' moving in a circle 'outside the outermost orbit', he does not mean that *this* heaven either is or could come to be outside itself, but rather another one surrounding another world. So perhaps it should be said that it was shown in the foregoing that there was one middle and one extremity,[544] and that the extremity was the revolving body. So there could not be anything else revolving outside the extremity, given that there is no other middle. Consequently it is not possible for the nature of the extremity to abandon this place and come to be in another place, so that not every extremity was here,[545] but [some] would also be there.

Alexander understands 'give up its own place' not as referring to the *nature* of the revolving [body] and the extremity, but to *this* revolving [body], and says that it can neither be nor come to be in any place other than the one where it is. However, as I said, it is not to the present purpose to show that *this* heaven is not outside itself, but rather that outside this outermost revolving body there is no other encircling body, in the same way as he has shown that there is nothing else either light or heavy outside this world,[546] but not that those in it are not outside it.

That there is nothing heavy or light, or in general any of the things which move in straight lines, outside the heaven, he shows by employing once again the proper method of demonstration.[547] For if there are [such things], they are so either naturally or unnaturally. For in either case they are of a nature to be in place. [But they are not there naturally.][548] For their natural places, to which each of them moves
283,1 naturally, were shown to be elsewhere, namely in this world, if indeed it was shown that there is one middle and one extremity.[549]

But if one were to say that one of the rectilinear bodies were there unnaturally, it is clear that those places will be natural for other bodies, since what is an unnatural place for one thing is invariably natural for something else.[550] But there is no other simple body besides these; and so neither will there be a place in which some other simple body can be unnaturally. And if no simple body can be outside this heaven, neither could any mixed body; for the mixed [bodies] are put together from the simple ones, and where there are mixed [bodies], there must also be simple ones.

One should note that he advances all of this demonstration as dependent upon things demonstrated earlier, when he showed that each body's natural place was not only single in form, but numerically one as well, and [located] in this world; and for this reason he says that it has been demonstrated that the revolving body cannot give up its own place, and that the proper places of the rectilinearly-moving [bodies] are here. 'Since other places are proper for them' makes this clear.

279a2-11 But in fact it is not possible [for any body] to come to be there. [For it will be so either naturally or unnaturally, and

> will be either simple or mixed. Consequently, the same argument will apply once more, since it makes no difference to consider whether it is, or whether it can come to be. So it is obvious from what has been said that no bodily mass either is or can come to be outside. For all the world is made from all of its proper matter, since its matter is natural and perceptible body. Consequently there are not now many heavens, nor are many coming to be, nor could many come to be:] this heaven is single, unique, and complete.

Having shown that there is no body outside the heaven he next demonstrates that one could not come to be [there]. For this was most particularly what was principally at issue, namely that not only is the world single, but that it is impossible for more of them to come to be, which he shows not only through there not being any body outside this world, but also through its being impossible for any to come to be [there].

That 'it is not possible [for any body] to come to be there', he shows by the same means. For it could be neither simple nor composite, neither revolving nor rectilinearly-moving, since there is one middle and one extremity, and each natural thing is not only single in form but numerically one as well.

Alexander says that he has supplied as the reason why no perceptible body either is or could come to be outside this world the fact that it is [made] of all the proper matter.[551] But perhaps he has done exactly the opposite, giving as the reason why this cosmos is [made] of all the matter there is [the fact that] there is no body, whether simple or composite, outside it:[552] for having resolved the difficulty which furnishes [the possibility of] there being many worlds, on the grounds that this world is composed of all the matter there is, he says that it remains to show this, namely that [it is composed] of all [of it], which he shows from [the fact that] there neither is nor can come to be any body outside the heaven.

The latter, then, is the cause of the former, given that demonstration [is effected] by way of the cause.[553] For even if the fact that it is [composed] of all the matter there is and the fact that there is nothing left over outside are mutually entailing, that there is nothing left over still seems to be prior to the fact that it is [composed] of everything: for we define the universe in terms of there being nothing left over.

He rightly says that 'it makes no difference to consider whether it is, or whether it is capable of coming to be': for things which are capable of coming to be, and most particularly things which are natural and are not impeded, *will* come to be, or else they possess the capacity in vain. But things which are capable of coming to be in virtue of choice frequently do not come to be, if the choice is altered or if they are prevented in some way. Alexander treats 'all the world' as being significant: for he says that speaking of the world as 'all' is a sign of its being [made] from all the matter there is; for one does not speak in this way of the perceptible man as 'all', nor of any other perceptible thing, besides the world.

> **279a11-18** At the same time it is clear that there can be neither place, [void, nor time beyond the heavens. For it is possible for body to exist in every place, while they call 'void' that in which there is no body, but in which it is possible that one will come to be. And time is the number of change, and there is no change without natural body. And it has been shown that there neither is nor can be body outside the heaven. Therefore it is obvious that there is] neither place nor void nor time outside it.

Having shown that there is no body outside the heaven, he proves on the basis of this that there are neither place, void, nor time there, deducing the conclusions by way of the second figure. In every place it is possible for there to be body; outside the heaven it is not possible for there to be body; therefore outside the heaven there is no place.[554]

And again: if void is that in which there is no body but one can come to be; but outside the heaven not only is there no body, but none can come to be in it; then there is no void outside the heaven.[555]

And again a third: time, being the number of change,[556] invariably exists wherever there is motion, and motion wherever there is a moving body; so if there is time wherever there is body too, and outside the heaven there neither is, nor can come to be, any body, therefore outside the heaven there neither is, nor can come to be, time.[557]

The Stoics, however, wanted there to be void outside the heaven,[558] and they established this through the following supposition. Suppose, they say, someone stands at the extremity of the sphere of the fixed [stars] and extends his hand upwards. If he does extend it, they infer that there is something outside the heaven into which he extends it; but if he cannot extend it, there will be something outside in such a way as to prevent his hand's extension.[559] And if he then stands at the limit of this and extends, the same question [recurs]. For something will be shown to exist outside this as well.

'That this is unsound', Alexander says, 'may be shown thus. If the world is the universe, and there is nothing outside the universe, there will be nothing outside the world. For it would not be the universe if there were something else outside it. Consequently no one would be able to extend their hand having arrived at the extremity of the heaven. Further, it is not at all possible and contrary to supposition for something to come to be there. For the divine body that moves in a circle is unaffectible and incapable of admitting anything of such a kind within itself.'

But this objection is perhaps not conclusive, since we often suppose impossibilities for the sake of the subsequent consequence of the argument.[560] Plato[561] indeed supposed someone to be standing on the fire-sphere where the fire exists in its purest nature, having his head inclined towards the earth, and having the ability to abstract some of the fire and place it in the scales and drag the balance towards the air. And although this is impossible, he supposed it in order to show that down and up, heavy and light, do not exist by nature at the sphere. Indeed he concludes from the supposition that 'when two things are raised at the same time by a single force, the smaller necessarily yields more and the larger less to what constrains them by force, and the former is called heavy and downward-moving, and the smaller light and upward-moving'.[562] And Aristotle in the fourth book[563] will suppose the earth to be transferred to the place of fire.

But what ought rather to be said, as Alexander himself indicated, is that if the world is the universe, and there is nothing outside the universe, the supposition would be the same as if he were to try to extend his hand into the non-existent. For if he extends it, there will be a place that receives it (and it will no longer be non-existent); but if he cannot extend it, something must prevent it. So the supposition which provides what is sought in the imaginary case, that there is something outside the universe, whether void or solid, is absurd.[564]

'And', says Alexander, 'from the very supposition of the void they destroy the reality of the void. For let there be, if it were possible, void outside the world. This must then be either finite or infinite. But if it is finite, it must be bounded by something, and the same argument recurs at the limit of the void: someone either extends or does not extend his hand. For what will they say? If it is infinite, as Chrysippus believes, they must say that there is an infinite stretch of void, which, while capable of receiving body, does not receive any; but of relatives it is necessary that if one exists the other does too,[565] so if something is capable of receiving, there either is or could be something which is capable of being received. But these people say that there is no 286,1
infinite body which is able to be received by the infinite void. Therefore it cannot be capable of receiving it either.'[566]

Xenarchus[567] altered 'capable of receiving' to 'receptive', as if in this manner to resolve the absurdity deriving from relations in the supposition; but certainly the alteration did not accomplish anything more. For the receptive is nothing other than that which is capable of receiving, and being of such a kind it remains relative.[568] 'Moreover', he [sc. Alexander] says, 'if the world were in an infinite void, what would be the reason for its staying in the [place] where it is? For no body which has any inclination can stay still without something external preventing it [from moving], unless it is in its proper place. But there is no difference in the void which could make this [part of the void] proper for the body in it but that one not. But if the world is not at rest but moving, why should it move here rather than there, given that the void is undifferentiated? But if moved in all directions, it would be dispersed. For God could not plausibly himself be resisting them: for being fire, he himself should move with fire's motion.'[569]

Or would they say against Alexander that 'we do not suppose that fire is of such a kind[570] when we suppose it to be immortal?'

He [sc. Alexander] says: 'to hold that the world depends for its cohesion upon the *pneuma*[571] that is spread throughout it, where there is void, is to talk vacuously. For each of the other bodies, which are themselves held together by *pneuma*, ought not to move position. Furthermore the *pneuma* would contribute to preventing the dispersal of the parts of what it is in by holding it together, but not however to not being moved, being itself mobile and an impediment to the natural inclination of no body; in addition to which it is false to say that each body is held together by a certain *pneuma*, as we have shown elsewhere.'[572]

Alexander also attacks those who consider what has been said to be true, following the imaginary case of the hand.[573] 'For', he says, 'we imagine many impossible things; and each person may imagine himself outside the town, and many times his size, or much smaller and the size of a millet-seed.'

So much for these issues. Plato supplied two reasons for there only being one world, one of them derived from the exemplar: 'for', he says, 'in order that the world might be uniquely similar to the perfect animal, the maker made neither two nor an infinite number of worlds, but there is and will be only this single, only-begotten heaven.'[574] For if it were one among others, this one would not be perfect among perceptible things as the exemplar is among intelligibles, and nor would the exemplar of the world be single, given that, as I tried to
287,1 show a little while ago,[575] there is one image for one exemplar, towards which it is immediately and pre-eminently likened.

The other reason Plato gives for there being one world, the one which Aristotle welcomed, was that it was composed of all the natural body there was. In order to make plain the harmony between Aristotle and his teacher in these matters, which extends even as far as terminology, I will set these words of Plato down for comparison: 'the composition of the world exhausted each of the four elements entirely, for the composer composed it of all the fire and water and air and earth, leaving out no part or power of any of them, with the intention first of all that the animal should be most particularly a perfect whole of perfect parts, and in addition that it should be single, with nothing left over from which another like it might come to be, and further that it be ageless and free from disease.'[576]

Then, having added a few things, he concludes: 'on account of this reason and calculation he made this single whole from the whole of everything, complete and ageless and free from disease'.[577] Aristotle says the same thing in the same words, that being [composed] of all of its proper matter the world is one and unique and this heaven is complete.

> **279a18-b3** For this reason the things which are of a nature to be there are not in place, [nor can time cause them to grow old, nor is there any change for any of the things which are placed above the outermost movement, but unalterable and unaffectible, being possessed of the best and most self-sufficient of lives, they persist throughout all the ages. Indeed this name was uttered by the ancients in a divine manner: for the end which encompasses the time of the life of each thing, beyond which there is none natural, is called its age. And by the same reasoning the end of the whole heaven and the end which encompasses the whole of time and infinitude is an age, deriving its name from its always being, [and so is] immortal and divine. The being and life for everything else depends on this; for some things more precisely, for others more obscurely.[578] And indeed, just as in the non-specialist philosophical works, it is often made perfectly clear by the arguments regarding divine things that every first and highest divinity must be unchangeable;[579] this being the case, it confirms what has been said. For neither is there anything else greater which will move it in any way (since that would then be more divine), nor is there anything bad about it, nor does it lack any of its proper beauty. So it is reasonable that it move with unceasing motion. For everything stops moving when it arrives at its proper place, while for the body which moves in a circle the place] from which it starts is also the place at which it ends.

Alexander says that what is said here is said either of the Prime Mover, which seems to be outside all bodies (since it is not in any of them) and not in place (for it is incorporeal), or of the sphere of the fixed [stars]. And he interprets everything up to 'while for the body which moves in a circle the place from which it starts is also the place at which it ends' as being rather about the latter.

For (he says) he has shown in the *Physics* that it is not in place,[580] since place is the limit of the encompassing body,[581] and it is not encompassed by any other body. Those things which time encompasses are in time, as existing in a part of time. So if there is no body outside it nor time encompassing its being, it will neither be in place, nor grow old under the influence of time. And this fits in with what Plato said: for he says that 'he made it complete, and ageless, and free of disease'.[582]

If by 'above the outermost movement' he were speaking about the 288,1
first cause (Alexander says), he would be referring to [the region]

above the orbit of the sphere of the fixed [stars]; while if he were saying these things about the divine body, the 'outermost movement' would mean the furthest of the rectilinear motions. For he is accustomed to call the motions of [things which move] rectilinearly the 'movements' and the circular [motion] the 'orbit'. So above the furthest movement there is all of the revolving body, which he says is neither in place nor in time, being eternal and ageless. For while the divine body as a whole is not in place, parts of it are in place: the spheres of the planets are in place.[583]

After saying that things divine are 'possessed of the best and most self-sufficient of lives', and that 'they persist throughout all the ages', he wishes also to establish their immortality and eternality on the basis of the word 'age'; and having first articulated the full significance of [the word] 'age', he goes on from this to what is most properly meant by it. For we call the complete and all-embracing time of the life of something its age, 'beyond which there is none natural':[584] 'man, you die young from the ages', says Homer,[585] i.e. before the completion of your naturally-allotted time.

Aristotle says that 'age' more properly means the wholeness and completeness which gathers together the infinite time of the whole heaven. For this rightly derives its name 'from its always being',[586] and it is 'immortal and divine', and it is the cause of immortality and eternality for those things which immediately and properly participate in it, and, by possessing and comprehending the whole of time within itself, it assigns a partial extension of life to the other things. For 'being and life' are assigned by it to everything, 'for some things more precisely, for others more obscurely'.[587]

And Aristotle understood 'age' prior to this, in the *Metaphysics*,[588] as being the internal capacity of the primary mind, which Plato calls 'an exemplar' of time.[589] For while this remains single in number, an ageless image of time is produced from it. And he says that the being of everything depends upon the heavenly age, since its motion is the cause of being for everything in the world of generation.

That the divine is eternal, he says, is attested by 'the non-specialist philosophical works', in which 'it is often made perfectly clear' in arguments that the divine is necessarily changeless and in every way the first and highest being; and if changeless, then eternal. By 'the non-specialist philosophical works', he means those which were in the
289,1 original arrangement addressed to lay-people, which we usually call 'exoteric', just as [we call] the more serious ones 'lectures' and 'treatises'.[590]

He talks of this in *On Philosophy*.[591] For in general, wherever there is a better there is also a best. So since, among existing things, one thing is better than another, therefore there will be something best, which is the divine. So what changes is changed either by something else or by itself, and if by something else, by something either better or worse, and if by itself either towards something worse or striving towards something finer. But the divine has nothing better than itself by which to be changed (since that would then be more divine); and nor is it right for the better to be affected by the worse; and if it were [affected] by the worse, it would acquire something bad, but nothing in it is bad. But nor can it change itself by way of desiring something finer, since it is lacking in none of its own finenesses. And it does not change for the worse, since not even man makes himself worse voluntarily, nor does it possess anything bad which it would acquire from a change for the worse.[592]

Aristotle adopts this demonstration from the second book of Plato's *Republic*:[593] 'is it not necessary, if he abandons his own form in some way, that he is changed either by himself or by something else? – It is necessary. – But the best things are least likely to be altered and changed by something else, as the body is by food and drink and labour.' Then having shown this, he argues: 'but would he change and alter himself? – Clearly he would (he said) if he alters at all. – So would he change himself towards something better and finer, or towards something worse and meaner than himself? – Necessarily (he said) towards something worse, if he were to alter at all. For we will not at any event say that God is lacking in any way in fineness or excellence.'[594] And having shown that no one makes himself worse voluntarily, he concludes: 'but, so it appears, if each of them is as far as possible the finest and best, it will remain in its own form.'[595]

And he [sc. Aristotle] plausibly says that its motion is eternal as well. For 'everything that moves naturally[596] stops moving when it gets to its proper place' from some other, and there is a beginning for their motion when they are in a foreign place, and an end to it when they have come to be in their proper one. But for the body which moves in a circle the same place is both beginning and end for its motion. And because it is always at its end and in its proper place it is always in its proper good, while because it is always at its beginning, its motion is unceasing. For nothing which is at its beginning ceases moving.[597]

All of what I have just been saying, from the beginning of the 290,1
discussion to the end, Alexander prefers to interpret as applying to the revolving body. But the more recent interpreters urge that everything which has been said should be understood as applying to the immobile causes[598] which move the heavenly bodies. That not all of it can be interpreted as applying to the heavenly body is clear, I think, from what has just been said. For having shown that there is no body, either simple or composite, outside the heaven, he argued 'at the same time it is clear that there can be neither place, void, nor time beyond the heavens'.[599] And having shown this and concluded 'therefore it is obvious that there is neither place nor void nor time outside',[600] he then infers, as it were, from what has been said some corollaries: 'for

this reason the things which are of a nature to be there are not in place, nor can time cause them to grow old',[601] and so on, by 'the things there' clearly meaning those outside the heaven. So how could the heaven be said to be outside the heaven? And how could he say that there is no change of any kind for any of the heavenly bodies, when he sees that their change of place is unceasing? And neither does he say that the things which are positioned above the furthest movement are heavenly bodies.[602]

Moreover he speaks frequently of 'being moved in a circle',[603] and not invariably of 'revolving'. Indeed he says in what has gone before '[there is] another [place] for the body that is moved in a circle, which is the extremity'.[604] And how could he call the heaven 'the first and highest divinity'? How could there not be something else greater than the heaven to move it, given that the immobile cause, which is the cause of its eternal motion, is greater?

But that not everything can be interpreted as being about the intelligible causes is clear from 'indeed[605] by the same reasoning the end of the whole heaven'. Furthermore, the considerations at the end plausibly show that the heaven moves with an unceasing motion. But the passage is self-evidently obscure, otherwise excellent men would not be in such dispute about it. But perhaps these things are said concerning the intelligibles as being above the bodily nature of things, and they are said in this sense to be outside it. For whenever he inquires if there is something outside the heaven, the 'outside' does not signify some place.

And having offered as the most important meaning of 'age' for us that which comprehends the time of each thing's life, he proceeds from that to the heavenly life, saying 'and[606] by the same reasoning the end of the whole heaven', and for this reason here was referring to the heaven. Then [he moves] from this to the age beyond the world, when he says: 'the end which comprehends the whole of time and infinitude is an age'.[607] And this is indeed its most proper usage: that upon which
291,1 both being and life depend,[608] more precisely for the heavenly [bodies], more obscurely for those beneath the moon.[609] And in bringing up in testimony the changelessness of everything divine which is first and highest, he seems to be speaking of the intelligible and immobile principles which move the heavenly spheres.

That he does speak of many of them is made clear by the 'every',[610] that it is unmoved by 'unchangeable', and that it is beyond the heaven by 'first and highest'.[611] But that there is nothing greater which moves it is more appropriate to it rather than to the heaven, since the immobile cause which moves the heaven is greater than it.[612] And if indeed 'nor is there any change for any of the things which are placed above the outermost movement' does refer to the intelligibles, he says that this is confirmed by what is 'often made perfectly clear in the non-specialist philosophical works', namely 'that every first and highest divinity must be unchangeable'.[613]

But 'so it is reasonable that it move with unceasing motion' (if indeed that is what was written)[614] is clearly said about the heaven; and it seems to be connected with what went before, as being about the same thing, which said that nothing of the motion of the thing which moves unceasingly belongs to it.[615] This seems to be said about the same thing; but how could it be said about the same thing, if the former was shown to be unchangeable and such that there was nothing 'else greater which will move it in any way', while for the latter it is reasonable that it move with unceasing motion?

And if someone thinks that 'unchangeable' means with respect to changes other than of location, I would be astounded if it would be [consistent with] Aristotle's precision to say of things which change location that there is no change of any kind for any of them. But it is self-evident that the expression 'so it is reasonable that it move with unceasing motion', and what follows, refers particularly to the creator and the prior things concerning the heavenly bodies.[616]

But if the text is such as I have found in some transcriptions 'so it is reasonable that it *cause*[617] an unceasing motion' (and not 'move with', as Alexander thought), then this and everything before it can without violence refer to the intelligible and immobile causes, and the things said to be outside the heaven as transcendent. For since these things are unchangeable and immobile, for this reason the heavenly bodies are moved primarily by them with unceasing motion. For while the mover is always similarly disposed towards the same things,[618] that which is primarily moved by it is connected with it by its superior suitability,[619] and is moved unceasingly.

Cyclical motion was shown in Book 8 of the *Physics*[620] to be the only continuous and eternal [motion]. And because the heaven moves in a circle emulating mind, desiring its particular nature, and participating in an actuality which is always similarly disposed towards the same things, for this reason it moves with unceasing motion. 292,1

For the things which move in a straight line, having one beginning for their motion and another end, move from their beginning and rest when they arrive at their end and proper place. But, for the body which moves in a circle, since the place from which it begins and to which it ends is the same, it is always in its end because it is always in its proper place, and it is always in its beginning because it is always moving towards the end.[621]

Notes

1. The lemmata are given in Simplicius in abbreviated form: square brackets mark off those portions of Aristotle's text which are not reproduced in the MSS of Simplicius, or in Heiberg.

2. *Cael.* 1.2, 268b11-13.

3. i.e. the arguments of *Cael.* 1.2-4 which establish the existence of the fifth body, the ether, the element of the heavenly bodies distinct from the sublunary four of earth, water, air and fire.

4. See Simplicius *in Cael.* Prologue, 1,2-24; cf. 2,25-3,12. Much of the Prologue to Simplicius' commentary deals with the question of Aristotle's primary purpose in writing *Cael.* Alexander thinks that its subject-matter is, absolutely generally, the universe as a whole; Iamblichus holds that the subject properly speaking is the heavens, and that Aristotle treats of the sublunary elements in it only insofar as they are dependent upon it (1,24-2,5); Syrianus and his followers say that the subject really is the heavenly bodies, and anything said about the sublunary elements in it is said only insofar as it contributes to an understanding of them (2,5-16). Simplicius' own view (4,25-5,4) is that the sublunary elements are treated of in their own right, but as essentially subordinate to the heavenly element, hence the title of the treatise.

5. i.e. the subsequent proofs of the finitude of the cosmos.

6. Prologue, 3,12-6.7; see Hankinson, 2002.

7. And not, as Alexander held, in the interest of giving an account of the universe as a whole as such: n. 4 above.

8. i.e. as an exhaustive disjunction.

9. Theophrastus in Simplicius *in Phys.* 1.2, *CAG* IX 24,26-25,1 = 13 A 5 DK; Hippolytus, *Refutation of All Heresies* 1.7.1 = 13 A 7 DK.

10. Theophrastus in Simplicius *in Phys.* 1.2, 24,13-25 = 12 A 9 DK; Hippolytus *Refutation* 1.6.1 = 12 A 11 DK; scholars disagree as to whether Anaximander's *apeiron* was literally infinite in extent, and also as to whether the infinity of worlds were successive or concurrent: for discussion, see Kirk, Raven and Schofield, 1983, 109-17, 122-6.

11. See Diogenes Laertius 9.31 = 67 A 1 DK; Hippolytus *Refutation* 1.13.2 = 68 A 40; and see 1.7, 242,14-26 (= 67 A 14 DK) below.

12. 'Opposition' = *enantiôsis*. The pun is difficult to reproduce in English: in the first sense it means 'dispute', in the second 'opposite condition'.

13. Simplicius *in Phys.* 1.4, 155,30-156,1 = 59 B 2; 1.4, 179,3-6 = 59 B 15-16 DK.

14. The main Presocratic exponent of this position was Empedocles (31 B 8-12 DK); but it is also the view of Aristotle (*Phys.* 1.6-8; and cf. e.g. *Gen. Corr.* 2.2-5); and in a sense goes back to Anaximander (Aristotle *Phys.* 1.4 = 12 A 9 DK); Theophrastus in Simplicius *in Phys.* 1.2, 24,13-25 = 12 A 9 DK.

15. *epagôgê*: strictly speaking enumeration of cases rather than induction proper, since the term in Aristotle's Greek need not refer to any inferential procedure; but 'induction' is a translation hallowed by usage.

16. Presumably a reference to Epicurus and his followers: see n. 17 below.

17. The 'smallest magnitudes' are distinct from the atoms themselves; rather they are the minimal spatial units that make up the atoms (see Epicurus, *Letter to Herodotus* [*ad Hdt.*] 56-9), Lucretius' *minima* (1.599-634, 746-52); see Long and Sedley, 1987 i, ch. 9.

18. The continuousness, or potentially infinite divisibility, of space and spatial magnitudes is a cornerstone of Aristotle's physics (*Phys.* 3.6, 206a9-29; 6.1-2); if space is quantized, however, as the atomists have it, then no line made up of an odd number of quanta can be bisected; moreover, no triangle (such as the right-angled isoceles) where one side is incommensurable with the others (i.e. where its length is expressed by an irrational number by comparison with the others, such as √2) can be constructed. 'It is not difficult to refute the theory of indivisible lines', Aristotle says breezily (*Phys.* 3.6, 206a16); see ps.-Arist. *On Indivisible Lines.*

19. i.e. as opposed to simply enumerating examples: 202,25-203,2.

20. Simplicius makes rather a meal here out of what is merely an Aristotelian jocularity – if mistakes about the smallest things make a huge difference, then mistakes about the infinite will make even more of one. But of course mistakes about large things are not necessarily large mistakes.

21. *arkhê*, here in the sense of natural principle or element: cf. Aristotle *Phys.* 1.2, 194b15-21.

22. Or, as we would say, *reductio ad absurdum*; see below, 208,27-8; 1.6, 217,22, 222,34.

23. *proslêpsis*, literally additional assumption; in hypothetical syllogistic a technical term for the categorical or minor premiss.

24. *Cael.* 1.5, 272a5, below. At *in Cael.* 1.2, 12,10-11, Simplicius describes the proposition that the heavens move in a circle as one of the six 'hypotheses' upon which Aristotle bases his argument in *Cael.* 1.2-4 for the existence of the ether (see Hankinson, 2002, ad loc.; and Hankinson, forthcoming 1), and says there that perception confirms it. Even so, it was not obvious to all the ancients: Xenophanes thought that the apparently circular path of the sun was a trick of perspective, and that it actually moved in a perfectly straight line across the sky, there being a new one every day (Aëtius 2.24.9). Aristotle offers arguments for the position that the heavens rotate at *Cael.* 2.4; and Ptolemy gives a set of arguments, some of them derived from Aristotle, for the view, as well as refutations of the alternatives: *Syntaxis* 1.3; see Taub, 1993, 45-60.

25. *sunêmmenon*: the Stoic technical term for the conditional, which became standard in later Greek logic (cf. e.g. Sextus Empiricus *Outlines of Pyrrhonism* [*PH*] 2.110-12); the conditional in question is that of 204,4-5: if the heavenly body were infinite in magnitude, it could not move in a circle. For the expression 'proving the conditional', see *in Cael.* 1.6, 219,11-13.

26. This gloss of Simplicius' is not unexceptionable; and it seems preferable to suppose that by 'interval' here Aristotle means the linear distance between the two end-points of the produced radii, whether measured as a straight line or as an arc of the circumference of the circle of which they are radii (earlier Aristotle has claimed that distances on a circumference are properly measured by the chord which links their termini: *Cael.* 1.4, 271a10-13; cf. *in Cael.* 1.4, 147,23-148,26); see Hankinson and Matthen, forthcoming, ad loc.

27. i.e. the greatest possible magnitude which is still contained within the projected radii.

28. It is hard to see how this qualifies as a characteristic of infinite intervals only, at least if the interval is taken to be the enclosed area.

29. The difficulty with understanding 'interval' to mean 'enclosed area' is brought out by the phrase 'touching them'; Simplicius assumes that the hypothesized extension is outwards from the centre, i.e. we are to imagine the area being pushed over the edge of the infinite circumference – but then it is hard to see how such an area could be 'touching' the lines in any sense. On the alternative interpretation, the line connecting the two extreme points of the infinite radii is such as to have nothing 'beyond it ... which touches the lines', in the sense of their being no greater length which can be contained within the radii; 'touches' here then means 'makes contact with but does not exceed'.

30. Here Alexander makes the (Aristotelian: *Phys.* 3.5, 204a20-34) mistake of supposing that no infinite magnitude can be a proper subset of any other (infinite) magnitude: see below, nn. 39, 41-2.

31. i.e. the fact that the lines projected from the centre are infinite; the dispute between Simplicius and Alexander here is more than usually trivial and scholastic.

32. All surviving MSS of Aristotle read 'Further, it is always possible' here: see nn. 34-5 below.

33. i.e. in the series of numbers that tend towards infinity there is no last member.

34. None of the surviving MSS of Aristotle reproduce the reading 'if'.

35. i.e. Aristotle would not need to write 'always' here, since the generality of any such claim in geometrical contexts is self-evident. In spite of Simplicius' contentions here, modern editors all read 'always' instead of 'if' here, along with the MSS. The sense is not affected.

36. The words in [ ] are not included by Simplicius – but clearly they belong to this lemma and not the subsequent one.

37. See n. 24 above.

38. *Cael.* 1.2.

39. Here Alexander commits another common ancient mistake (one also made by Aristotle: *Phys.* 3.5, 204b19-22) of supposing that something infinite in extent must exhaust all the available space: see above, n. 30.

40. i.e. the concepts cannot apply to infinite magnitudes; cf. nn. 42, 50 below.

41. i.e. there will be some determinate measure (finite length) m such that each of the two magnitudes will be $n.m$; but then, as the next sentence points out, they will be finite: see further 208,2-4, and n. 49 below.

42. i.e. if the infinite exceeds the finite by an infinite amount, there will be two bodies (the original infinite body, and the infinite surplus) both of which are infinites, but of which one is greater than the other – and this is supposed to be an absurdity (Aristotle thought it was too, which is why he held there could be no actualized infinity: *Phys.* 3.5, 204a20-34: above, n. 30). It was not until the work of Dedekind in the nineteenth century that mathematicians succeeded in divorcing the concepts of size and proper containment: indeed, this 'absurdity' becomes part of the Dedekindian definition of an infinite set, as one which has a proper subset of itself of the same size as itself.

43. *Phys.* 3.5, 204b28-9, 205a23-5: an infinite element would overwhelm any other finite ones.

44. The rest of the discussion down to 108,8 is not marked as a quotation from Alexander in Heiberg's text, probably rightly (there are no markers of oratio recta – 'he said', etc. – within it); but it is probably at least a very close paraphrase of Alexander's text. Here Alexander supplies the specific argument regarding the case of the 'cyclical and spherical body', after the 'general argument' regarding any infinite body of 206,22.

45. i.e. the notion of equality cannot make sense in the case of infinite quantities since by definition there are no multiples of measures in virtue of which they can be equal; compare the case of 'double and half': 206,32-3 above.

46. One might object that a line might be infinite in one direction only, and hence 'come to an end at some point' (as the set of positive integers, taken in descending order, comes to an end at 1); but the point is that we are considering radii: definitionally, radii start from a point (the centre), and the issue is whether they can be infinite – if they can, then their production from the centre indeed cannot come to an end. On the 'infinite in one direction only', see 272a12, and nn. 79, 81 below.

47. i.e. if their intervals are finite they are finite also, and if their intervals are infinite, they will be infinite (this is of course true only on the assumption that the number of the intervals is itself finite in each case.

48. This sentence is rather garbled; the sense, as the succeeding lines show, should be: 'if the radii are to produce an infinite diameter', but it is impossible to glean that without wholesale alteration to the Greek, and I have preferred to leave Simplicius' text as it is.

49. i.e. if they are equal, they must be finite (because they are measured by equal measures: above, 207,1-3; n. 41.

50. Since the diameter is double the radius, and both radius and diameter are infinite: cf. nn. 40-2 above.

51. Here ends the lengthy paraphrase of Alexander: see n. 44 above.

52. i.e. Aristotle here relies simply on empirical evidence for the heaven's circular nature, since he has not yet proved it, in the manner that Alexander just has.

53. *Phys.* 4.11-14, esp. 220b14-32; cf. the definition of time as 'the number [i.e. measure] of change' (*Phys.* 4.11, 220a24-6).

54. See especially *Phys.* 4.8, 215a24-216a23.

55. cf. *Cael.* 1.9, 279b1-3; *in Cael.* 1.9, 292,1-7.

56. This is not of course true in the case of pure geometry: a line may begin from a point and continue infinitely (the 'infinite in one direction only: see nn. 46, 79, 81); but Simplicius is here clearly talking in physical terms: if a motion has a beginning, then it must also have an end, since it must have a natural terminus (a proposition more thoroughly investigated in *Cael.* 1.10-12).

57. The better of MSS of Aristotle, followed by most editors, have C here; but Simplicius clearly read A (see 209,21-2), and in fact that makes slightly better overall sense; although the diagram can function illustratively either way, the segment of the line AC will perform no function in the illustration if the line rotates around C. Elders (1965, 104) suggests that C represents a point on the earth's surface and BB a tangent to it, which makes good illustrative sense (BB will the represent the observer's horizon: see n. 60 below); see n. 59 below.

58. On the assumption that C represents the point at which the line cuts the earth's surface (see n. 57 above), the line 'describes a circle in respect of C' in the sense that C is the point at which the line is observable in its passage to an observer on the earth's surface (however, see n. 60 below).

59. In other words, Simplicius takes C simply to specify an arbitrary point finitely distant from A on ACE in virtue of which it makes sense to describe ACE as describing a circle (see nn. 57-8 above).

60. Since BB is infinite in each direction, the infinite ACE will cut it as soon as it veers from being parallel with it, no matter how far distant it is from A (and the time of the cutting will be twelve hours); but then it will cut it at an infinite distance from the point at which BB is directly overhead to an observer on the earth (on the assumptions of Elders – above, n. 57 – the moment at which ACE cuts BB will be the moment when a heavenly body rises above the horizon; the finite interval of time, then, will be between rising and setting); and strictly speaking there will be no first point at which the cut occurs, and no last one either.

61. i.e. a temporal beginning and end.

62. Adopting the reading of D; the other MSS have 'AEC' here, which is clearly an error, although the error might be Simplicius'.

63. i.e. a finite one.

64. It is not clear why Simplicius says this. If BB were finite, the cutting of it would be irrelevant to the proof that ACE moves an infinite distance in a finite time, since ACE would only pass through it for a finite distance, and there would be a determinate first and last point of the cutting; see n. 60 above. Perhaps the thought is that, given that BB is at a finite distance from the centre, then no matter how short a finite segment of it you take, since ACE is produced to infinity, at the extreme it will travel an infinite distance – but this is a different point from the one Aristotle is making in this passage.

65. i.e. in extent.

66. See n. 60 above.

67. Phys. 6.7, 238b13-17.

68. i.e. that an infinite body can traverse an infinite distance in a finite time.

69. The idea seems to be that Aristotle should not here assume the impossibility of an infinite traverse in a finite time – rather that is what the argument sets out to prove, by showing that assuming such a traverse entails the impossibility that an infinite interval should have a beginning and an end (a first and last point of the cutting in the revolution – note that it is spatial, and not temporal, points which are at issue here).

70. i.e. the *Physics*: see nn. 67 above, 72 below. For this designation of the second half of the *Physics* (the first half being referred to as *On Principles*), see *in Cael.* 1.6, 226,19-21, and n. 57 ad loc.

71. *hupokeimena*: a *hupokeimenon* is not a mere assumption; rather it is a premiss which has been 'laid down' like a foundation, and which can be taken as provisionally established (cf. *Cael.* 1.3, 269b18, 270b2; 2.3, 286a20-2, 286a30-1; 2.8, 289b5-6: 2.14, 296a24-97a8).

72. *Phys.* 6.7, 238b13-17.

73. Aristotle's argument, here given by Simplicius in rather garbled paraphrase, goes roughly as follows: if something infinitely long can traverse a finite distance in a finite time, it must do so by going at some (determinate) speed; but since speed is a function of distance and time, whatever that speed may be, it must be the speed at which some determinate finite magnitude can traverse that distance in that time: hence the infinite will be equal to the finite, which is impossible.

74. It might be objected that, for example, the front end of an infinite train, if it were travelling at the eminently finite speed of 20 mph, could cover a finite distance, 10 miles, in a finite half an hour. But for Aristotle (cf. *Phys.* 6.7, 238a19-b19) for something to move a certain distance, it must cover that distance as a whole – the whole infinite train needs to have traversed the ten miles of track; and on the assumptions canvassed, that is indeed an impossibility. See Hankinson and Matthen, forthcoming, ad loc.

75. The Greek is not easy to construe – this 'since' (*epeidê*) seems redundant, and should perhaps be omitted; alternatively perhaps '*ou diapherei*' ('it makes no difference') or the like has dropped out immediately before it.

76. The claim is not that, in two separate cases, in one of which one line only is moving and in the other of which both are, there will be no difference in the time taken for the passage (although of course there might not be – if each moves at half the speed of the original single mover: below, 212,15-18); but rather, as the next sentence makes clear, that there is in either case only one moment of separation.

77. i.e. no matter how slowly the lines move relative to one another they will within some (finite) time be free of one another.

78. i.e. an arc of finite length set against the heavenly sphere: below, 212,34-213,1.

79. Simplicius knows of an alternative reading, *epi thatera* for *epi mêkos*, which yields the sense 'in one direction only' (below, 213,33-214,2); see nn. 46, 56 above, and 81 below.

80. This supplement, or something similar, is required for the sense; alternatively, one 'not' should be omitted (or an equivalent change made).

81. Simplicius seems to take Aristotle here to mean that a line, even if it is infinite, is still bounded, in the sense that it occupies only one dimension (the remarks about planes in the succeeding lines appear to confirm this; but cf. 213,22-3); and if we retain the reading of the MSS of Aristotle here (see above, n. 79) that is reasonable. But he knew of an alternative reading (213,34-214,2), which would have Aristotle talking of lines bounded at one end only (cf. *Cael.* 1,12, 283a10; *in Cael.* 1.12, 347,24-30, 348,3-6); and that, as he says below, makes better sense in the context. See Hankinson and Matthen, forthcoming, ad loc.

82. 'In respect of another part' is obscure: presumably it means 'in respect of some but not all dimensions'; at all events, there seems to be no suggestion of considering a plane infinite in one direction only (see n. 79 above).

83. That the heavens move in a circle is supposed to be evident to perception (*Cael.* 1.5, 272a5-6; 2.4, 287a11-12; cf. *in Cael.* 1.2, 12,10-11; and n. 24 above); it has also been established that the heavenly bodies are composed of an element whose natural tendency it is to move in a circle (*Cael.* 1.2-4). The sphericity of the heavens is established at *Cael.* 2.4, 2.7; that of the heavenly bodies at 2.11.

84. Other MSS of Aristotle read *hêi* in place of *hê* here, yielding the sense 'since it cuts it at F'.

85. The text preserved by Aristotle's MSS here, and which was available to Simplicius, is not lucid; but Simplicius' explanation (cf. 214,23 below) that CE really means AB, and that the point is that no matter how far the revolving line moves it will still be in the same case as AB, i.e. it will cut E (or, as Aristotle has it, EE), is hardly convincing. Rather, if the MSS are right, the point must be that, no matter how far it moves, there still will be a line from the centre that cuts EE somewhere – hence it will never get clear of it.

86. See n. 84 above.

87. Aristotle's official account of place (*Phys.* 4.1-5) defines a thing's place as 'the limit of the surrounding body' (*Phys.* 4.4, 212a2-14), a view more congenial, apparently, to Alexander's interpretation; it is doubtful whether the idea of a thing's place being a set of spatial co-ordinates co-extensive with the extension of the thing, which as Simplicius implies is a sophisticated one, can be made compatible with it (on these issues, see Sorabji, 1988, ch. 12, 209-11; and ch. 1). Aristotle seems simply to be imagining that there is coincident with its moving surface a stationary plane (either interior or exterior: Alexander effectively supposes that it must be inside the outermost body, but this does not seem necessary), and which serves to measure its motion. He need neither ask nor answer the question, presupposed by Alexander's interpretation, of what this coincident plane belongs to as the surface of a body (at *Phys.* 4.5, 212b11-19, Aristotle allows that the universe as a whole is not really in place, or is so only accidentally, in virtue of its parts being in place).

88. See nn. 83, 89.

89. One argument for the sphericity of the whole derives from the impossibility (on Aristotle's view: *Phys.* 4.6-9; *Cael.* 1.9) of there being a void – but if the outer shape of the rotating sphere of the fixed stars were non-uniform (around the axis of rotation: as Simplicius correctly points out here, the strictures on void are compatible with there being cylindrical or conical – or for that matter ellipsoid – universes), then there would need to be absolutely empty space to accommodate the protuberances in the rotation: *Cael.* 2.4, 287a11-22.

90. The MSS of Aristotle all read *dielêluthen*, 'goes through', for Simplicius' *perielêluthen*, 'goes round', here: the sense is not affected.

91. *Phys.* 6.7, 238b13-17; see also above, 208,19-32; 210,15-20.

92. See 215,3-14 above, on Simplicius' interpretative difference with Alexander.

93. See n. 90 above.

94. i.e. that none of the other simple bodies is infinite in magnitude.

95. *Cael.* 1.3, 270b26-7.

96. For a similar remark, see *in Cael.* 1.7, 228,15. Aristotle's theory of the natural directions of motion (up, down, around the centre) naturally yields three elements; and, as Simplicius suggests here, Aristotle will on occasion talk of there being three elements (*Cael.* 1.8, 277b13-17; 3.1, 298b6-8; cf. Simplicius, *in Cael.* Prologue, 1,18-24; 1.8, 272,11-12; 3.1, 555,7-12). But in his treatment of the properties of the sublunary world in *Cael.* 3-4, he is clearly committed (as he is elsewhere) to there being four sublunary elements. This difficulty was seized upon by Aristotle's opponents, notably Philoponus (*Against Aristotle on the Eternity of the World* frs 1-5; Wildberg, 41-5), as a hopeless inconsistency; but things are not as bad as that. Commenting on *Cael.* 1.3, 270b26-31, where Aristotle states that there cannot be more elements since there are only three natural motions, Simplicius remarks that 'the reason why the simple motions are three but the simple bodies five is that, in the case of the bodies which move in a straight line, there is one which is unqualifiedly heavy and one which is unqualifiedly light, which are contraries to one another, but there are also two intermediates which have a share in each of them, although more of one than the other' (*in Cael.* 1.3, 144,29-145,1). Moreover, the fact that Aristotle does not simply generate his elements *a priori* from his simple motions is a testament to the fundamental empiricism of his approach: see Hankinson, forthcoming 1.

97. Or possibly: 'one of whose places is above and the other below'.

98. Contrary opposites are defined by Aristotle as being 'those items which differ the most of things of the same type; items which differ the most in the same subject; items which differ the most in regard to the same capacity; or are such as to exhibit the greatest difference, either *simpliciter*, or generically, or specifically' (*Metaph.* 5.10, 1018a26-31); and cf. *Metaph.* 10.4, 1055a3-5: 'since things which differ from one another can do so either more or less, there is also such a thing as the greatest difference, and this I call "contrariety" (*enantiôsis*)'; on the use of the notion in this passage of Aristotle, see Hankinson and Matthen, forthcoming, ad loc.

99. i.e. if there is a furthest point, it is something definite; see further 217,4-5, 218,24 below.

100. i.e. if the direction upwards simply continues indefinitely, there is *ex hypothesi* no limit to it, and hence no determinate uppermost place, and hence no contrary place to the centre which could serve as a focus for natural motion in that direction.

101. Replacing Heiberg's question mark with a full stop. At *Cael.* 2.14, 297b17-23, Aristotle takes the fact that bodies always fall at right-angles to the earth to indicate its sphericity (see also 1.5, 213,31-3, and nn. 83, 89); here Simplicius seems to be gesturing towards an argument in the converse direction.

102. The translation deliberately preserves the ambiguity in Aristotle's Greek, where *gignesthai* may mean 'come to be' in the sense of 'be generated', or simply in the sense of arriving in a place. Simplicius, probably rightly, takes it in the latter sense (*contra*, e.g., Elders: 1965, 109). See Hankinson and Matthen, forthcoming, ad loc.

103. *Cael.* 1.5, 272a11-b17; 272b28-273a1; cf. *Phys.* 6.7, 238b13-17; and *in Cael.* 1.5, 209,21-213,7.

104. And hence finite: 273a13-14.

105. Since no body can traverse an infinite distance, a piece of fire (say) at the earth's surface could never reach its natural place at an infinitely distant extremity.

106. Presumably a reference to *Phys.* 8.9, 265a19-20: but there Aristotle actually says: 'for the impossible does not occur, and to pass through an infinity is impossible'; here the idea is that if a process is not in principle completable, there is no process at all, and it cannot even be said to have got under way: see 1.7, 229,17-20, and n. 210 ad loc.

107. Reading *anôthen katô*, with D, in place of Heiberg's *katô*; cf. 217,30-2 above; and esp. 1.7, 229,15-16.

108. i.e. we must assume not just that things can be in the middle, but that they can have come to be there, i.e. that it is possible for them to have journeyed there from their natural place.

109. *Cael.* 1.5, 271b28-272a7.

110. The infinite is characterized at *Cael.* 1.5, 271b33-272a2 as being such that you can always take more of it than any given quantity (i.e. for any finite *n*, it is greater than *n*); cf. *Phys.* 3.6, 206b33-207a2: 'it turns out that the infinite is the contrary of what people say: for the infinite is not that of which nothing is outside, but rather that of which something is always beyond.'

111. i.e. given that up and down are contrarily opposed places, and that contraries are the furthest apart in their domain, then since in an infinite distance there is no furthest point, contrariety of place cannot apply to infinite distances. But the most that arguments of this sort can show is that if there are contrary places (in this sense), they cannot be infinitely separated; and Aristotle's atomist opponents would of course deny precisely that there are contrary natural places.

112. See 216,29, 217,4-5 above.

113. i.e. by modus tollens: if *p* then *q*; but not-*q*; therefore not-*p*, which is indeed the form of the succeeding argument (219,6-11). Simplicius' language suggests that an earlier argument also took this form; but it is hard to see which he means, if he does indeed intend this implication (cf. 217,22 above, where he rightly describes an argument as a *reductio ad impossible*).

114. sc. naturally.

115. See 1.5, 204,8, above.

116. The argument for this does indeed involve most of the preceding lemma: 273a27-b15; crucially it depends on the principle that 'the magnitude is proportional to the weight', which, while intuitively plausible, is not a conceptual necessity; see Hankinson and Matthen, ad loc.

117. *ekthesis*: to prove a point by *ekthesis* is to take a typical case and generalize what follows from it since it is typical ('by exposing to mental view a particular instance of the class denoted by the middle term': Ross, 1949, 311): *An. Pr.* 1.6, 28a23, b14.

118. Aristotle conceives of the body, and its weight, as analogous to lines:

A——G——F——D——B
<—E—>
|——|——|——|
<—— C ——>

119. i.e. E will divide into C without remainder. On the notion of 'measuring out', see 1.5, 207,1-3: nn. 41, 49.

120. See n. 118 for diagrams. Assume an infinite body AB of finite weight C. Then take any finite sub-magnitude of AB (call it BD); BD's weight (call it E) will evidently be finite, and less than C. Hence, since C is finite, there will be some number n such that $E.n = C$ (the assumption actually canvassed here is that n will be a whole number, and hence that E will 'measure out' C; but this is unecessary, and the argument can be generalized: see 220,14-25 below for the 'incommensurable' case; and see 220,28-9). But then, on the assumption that the weight of any single elemental stuff increases in direct proportion to its size, there will be some finite magnitude $BF = BD.n$ of weight C; hence the infinite whole will have the same weight as a finite proper part of itself.

121. Presumably the sense here is 'if the bodies are of the same type', although Simplicius' mode of expression does not make this entirely clear. Strictly speaking, uniformity of density is irrelevant to the greater body/greater weight principle (as Simplicius notes below: 221,7-11), although not of course to the principles of proportion that Aristotle deploys in these arguments: see further, nn. 127-9 below.

122. See n. 118 above.

123. In the sense of 'more general'.

124. 'commensurable and incommensurable' not in the standard mathematical sense in which two magnitudes are incommensurable just in case their quotient is irrational, but in the sense rather of whether or not one may be divided into the other without remainder: see n. 120 above.

125. This proof relies on the so-called 'axiom of Archimedes'.

126. i.e. in terms of density. A 'uniform' (*homoiomeres*) body is one no part of which is distinct in form from any other part.

127. cf. n. 121 above; but this is not strictly speaking true: see n. 129 below.

128. Simplicius is assuming that the non-uniformly distributed mass either thins out or thickens up, depending upon which way you are going; but in either direction, you simply adjust the amount of stuff taken in the appropriate direction to equalize the weight; however, see n. 129 below.

129. i.e. on the supposition that the infinite body has a finite weight, then no matter how light each part of it is, by subtracting finite quantities of it one will eventually (but after a finite number of operations) arrive at a finite aggregate of the same weight as the original infinite body. However it should be noted that there is a way of arranging a finite mass such that it exhausts a space of infinite extent, by means of a Zenonian series: consider it arranged in a series of concentric shells each of depth d, and let the first shell be of weight w, the 2nd of weight $w/2$, the third of weight $w/4$ ...: the total will sum to $2w$, although it will be infinitely extended (cf. 1.7, 228,24: n. 201 ad loc.).

130. As Aristotle goes on to say: 274a17-19 (and cf. 272a25-7 above); see also 225,30-4 below. This equivalence is of course only intuitive for a positive conception of the property of lightness such as Aristotle's (thus, e.g., a greater mass of fire will exhibit a greater lightness than a smaller one: cf. *Cael.* 1.8, 277a33-b5; and see *in Cael.* 1.8, 267,10-28.

131. The text here is somewhat odd, and may be corrupt; but the sense is not in doubt.

132. This is true in Aristotle's physics only if other things are equal; the velocity of a moving body is also affected by the density of the medium through which it moves: 'we see the same weight or body moving faster than another for two reasons, either because there is a difference in what it moves through ... or because, other things being equal, the moving body differs from the other owing to excess of weight or lightness .... A then will move through B in time C, and through D (which is less dense) in time E (if the size of B equals that of D) in proportion to the density of the impeding body' (*Phys.* 4 8, 215a24-215b4). But such refinements involving the differential densities of media can safely be ignored in this context.

133. Literally 'following from the first'; but the second 'axiom' does not of course follow from the first in the sense of being logically entailed by it: rather, it represents a further specification of it.

134. These proportions may be represented by the following generalized principle:

[P] for any two objects O_1 and O_2, if O_1 is of weight w and O_2 of weight $w.n$, then if O_1 covers distance d in time t, O_2 covers d in t/n.

135. *Phys.* 6.2, 233a32-b15.

136. And hence that the time taken, even by the smallest (and hence slowest) thing is not infinite, no matter how great the distance covered: 222,27-30 below.

137. By principle [P]: n. 134 above.

138. *Phys.* 6.2, 233a32-b15.

139. See below, nn. 153, 156.

140. This translates the text as apparently read by Simplicus (224,10-14 below): *allo gar an ti peperasmenon elêphthê en tôi autôi logôi pros heteron meizon en hôi to apeiron hôst'en isôi khronôi*; the MSS of Aristotle all read: *allo gar an ti peperasmenon elêphthê en tôi autôi logôi en hôi to apeiron pros heteron meizon hôst'en isôi khronôi*; both are difficult to construe and to make sense of (see Guthrie, 1936, 50-1; Tricot, 1949, 25; Moraux, 1965, 20). If the text is read as in the MSS, punctuating with commas after *logôi* and before and after *meizon*, following Moraux, it yields the sense: 'for some other finite weight has been taken in the same ratio as that in which the infinite stands to the other, but larger than it, so that, etc.'; 'same ratio' must in this case mean 'same ratio determined by the temporal relations between the original finite and the infinite bodies'; this is clearly the best solution available retaining the MSS readings. But it is not clear that the Simplician text is not preferable (although it is equally unclear how Simplicius himself construed it: nn. 153-4 below). The only problem with it is that the 'some other' seems out of place, since the *ti peperasmenon*, on this construe, refers to the original finite body of 274a2-5; but note the interpretation of Alexander: 225,6-8 below. See further Hankinson and Matthen, forthcoming, ad loc.

141. That of two unequal weights, the greater will move the same distance in a shorter time than the lesser: above, 222,6-11.

142. See 222,8-9 above.

143. By principle [P]: n. 134 above.

144. Since if it did so it would have to move infinitely fast; i.e. it would have to cover any distance at all instantaneously – but that is impossible (n. 145 below).

145. i.e. there is no possibility of instantaneous covering of distance: cf. *Phys.* 4.8, 215a24-b22; 6.1, 231b18-232a18; 6.10, 240b30-241a6.

146. i.e. there need not be any continuous ratio linking weights to times; as long as we assume that the greater the weight the shorter the time, and that any movement must take some finite time, there will always be some finite weight (even if it cannot be directly determined by the application of proportional reasoning) which will move the distance in the same time as the infinite weight is alleged to; but see n. 129 above.

147. Aristotle assumes an objector arguing that, for any finite weight, no matter how big, and for its associated time of traverse of the distance, there will be a time which is shorter still, and hence we can assume that the infinite will traverse the distance in a lesser time than that of any determinate finite weight. The answer ought to be (and it seems that Simplicius is gesturing towards it here) that whatever time *is* taken, no matter how short, that will also be the time of transit of some determinate finite weight (and so the same absurdities will recur); but it is not clear that this is exactly what Aristotle says here.

148. Reading *diaretos* (*diareta*, Db, Heiberg; *diairet////* A; *diareton* B). For the arguments for the continuousness of time, along with that of space and physical magnitudes, see *Phys.* 6.1-2.

149. The view of the atomists, and of Strato: Sextus Empiricus, *Against the Professors* [*M*] 10.142-67; the former held space and body to be reducible to indivisibles too; for Strato, space and body were continuous (*M* 10.155).

150. i.e. *Phys.* 6.7: see ch. 5, n. 70. Simplicius (*in Phys.* 801,13-16) claims that 'Aristotle and his associates' refer to the first five books of our *Physics* as *On Principles* and the final three as *On Motion*; Porphyry, followed by Philoponus, treated *Phys.* 1-4 as *On Principles* and *Phys.* 5-8 as *On Motion*; and so does Simplicius at 226,19-20 below; as Ross (*Aristotle: Physics*, Oxford, 1936, 1-7) notes, the latter makes the better sense, although Aristotle's own habits of cross-reference are characteristically variable.

151. Again, this argument is persuasive only if the ratio of the increasing weight and the decreasing time is continuous; if on the other hand it is asymptotic (with every equal increment of weight, an exponentially decreasing contraction of the time), then the velocity of ever-increasing weights could converge upon a smallest time.

152. 'In place of' (*anti*) here might mean 'in the sense of': Simplicius would thus imply that the sense is potential here, even though the verb is in the indicative. But it might also mean 'rather than', i.e. by deliberate choice, and this is perhaps more consistent with the interpretation he proceeds to give (see n. 153). On the other hand, see below, 224,18: n. 157.

153. Something has apparently gone wrong with the logic of the argument here: the 'other finite magnitude' which 'might be taken' should be larger than the original finite weight, as Aristotle himself says (*Cael.* 1.6, 274a10-12; however, see Alexander's interpretation: 224,28-225,8 below, and n. 162); hence here it should be being compared to the original smaller weight; and this will get the ratios right. Let t_1 stand for the time taken by the original finite weight w_1 and t_2 for the time take by the infinite weight *W*; then there will be, given [P] (n. 134 above), another greater finite weight w_2, which is such that $t_1:t_2::w_1:w_2$. One could emend 'greater' to 'smaller', but the expression would still be slightly odd (why '*some* other weight'?); one might simply seclude *ti meizon heautou* ('greater in weight than itself'), adding *to* before *heteron*; but Simplicius clearly read something similar in Aristotle's text, and was trying to explain it. But Simplicius probably means by the 'some other finite magnitude' (224,11) which 'has been taken' (224,10) the original finite magnitude of *Cael.* 274a2-5. This, then, stands in the same ratio (determined by the ratio of the times) both to the infinite body and to some new finite body – hence the 'absurd' conclusion that something finite will move as fast as something infinite. However, if this is right, Simplicius inadvertently inverts then order of the ratio in the next clause (see n. 154 below); and there are problems making sense of the 'other' in 'a certain other finite weight has been taken', since this refers now to the original finite magnitude. It may just be a slip on Aristotle's part; or it may simply be 'other' than the infinite magnitude which has just been discussed. At any event, the problems are occasioned as much by the difficulties of Aristotle's text as by anything else: see n. 140 above.

154. If the interpretation of n. 153 is correct, and the 'certain finite body which has been taken' is the original one whose speed was compared with that of the infinite body in the preceding passage (274a2-9), then this expression of the relation between the ratios is, strictly speaking, the wrong way round: the original body (w_1) stands to the newly-assumed larger finite body (w_2) as the longer time does to the shorter (shortest) time taken by the infinite body: but even though the relation is one of inverse proportion, it is natural to compare the lesser to the greater in each case.

155. Above, 222,15-17 (see n. 134); 223,6-11, 16-18, 20-1.

156. If 'in place of' of 224,11 means 'in the sense of', then the option of taking 'it has been taken' 'more precisely' would mean that the verb had its normal indicative sense, referring to something that actually has already been assumed; but in any case, 'in place of' probably means 'rather than' (above, nn. 152-3), in which case that contrast cannot be intended here (since in the previous paragraph the sense was not potential). But if that is right, it is hard to see how *elêphthê* is being used 'more precisely' here than before. The thought might be that it 'has been assumed' already to stand in some ratio with the infinite (since their times of transit are commensurable) – and that is already absurd, independently of the conclusions that Aristotle draws about the other finite magnitude moving at the same speed as the infinite; this squares with what Simplicius says towards the end of the paragraph (224,26-7), but less well with the intervening text.

157. Again, the text is difficult and ambiguous; this seems the best supplement. The sentence might mean 'for if the infinite is supposed to move in the shortest time, in this very same time some other finite weight has been taken in the same ratio to the other greater [time] in which the infinite [moved]': but this clearly gives the wrong sense.

158. i.e. if there is a shortest time in which something covers a determinate distance d, then, if space is continuous, it should cover $d/2$ in half the shortest time – which is absurd.

159. sc. in all cases: i.e. there cannot be one smallest distance for one object and another for another.

160. i.e. it will move instantaneously.

161. Perhaps read *hapasan diastasin peperasmenên* for *hapasan peperasmenên*.

162. i.e. Alexander interprets the problematic passage as involving a second finite weight which is *smaller* than the original one of 274a2-9: above, nn. 140, 153.

163. Above, 274a10-11.

164. i.e. the second finite weight (w_2 of n. 153).

165. i.e. the first finite weight (w_1).

166. Simplicius' Greek literally runs 'by as many times (*hosaplasion*) as the distance which the finite weight moves is than the distance which the infinite weight moves in the same time, by so many times (*tosoutoplasion*) the weight which has been taken is than the infinite weight'. Normally one would understand 'greater' with both clauses – but this would give the wrong sense (the finite weight does not move further than the infinite weight in the same time). It is conceivable that the Greek might allow for the interpretation I have given in the text (effectively reversing which comparison is understood between the clauses), although I know of no other instance of this. Most probably Simplicius has simply lost control temporarily (although perhaps, in this passage, not uncharacteristically) of the logic of the ratios. But there is little doubt that this is what he wrote.

167. i.e. the weight taken, w_2.

168. Of the previous passage: 274a9.

169. See n. 150 above.

170. See n. 96 above.

171. *Phys.* 3.4-5.

172. Simplicius' practice in this regard is not consistent: see n. 150 above.

173. *Cael.* 1.5, 272a28-31.

174. 1.8, 276a18-277a12.

175. i.e. after the demonstration of the impossibility of there being any infinite body 'in more general terms' (274a28-9) which occupies Chapter 7.

176. *Phys.* 3.5.

177. *logikôs* ('reasonably', 'logically'): Aristotle contrasts arguments which are developed *logikôs*, i.e. those which progress from general conceptual considerations, with those which are *phusikôs*, drawn from the actual natures of things; in the case of the impossibility of there being an infinite body, the distinction is made at *Phys.* 3.5, 204a34-b10 (logical arguments); 204b10-206a8 (physical arguments); and see *Cael.* 1.7, 275b12, below. At *Gen. Corr.* 1.2, 316a10-4. Aristotle adds a third category, that of things which are *oikeia*, 'peculiar to' a certain subject-matter.

178. *Phys.* 3.5, 204b4-7.

179. *phusikôs*: i.e. from the natures of things; cf. nn. 112, 177 above.

180. *Phys.* 3.5, 204a20-6; 204b28-9; 205a23-5; and see *in Cael.* 1.5, 207,12-16, and n. 136.

181. *Phys.* 3.5, 204b10-19.

182. cf. *Phys.* 3.5, 204b22-35.

183. At first sight, this is a strange claim on Simplicius' part, since the sources uniformly credit Anaximenes with rejecting Anaximander's suggestion (12 A 1, 9-11 DK) of some indefinite stuff underlying all elemental change, the *apeiron*, in favour of postulating a single element, namely air (13 A 1, 4, 6, 7(1) DK). However he allegedly did so because air is 'close to the incorporeal' (13 B 3 DK): and Simplicius may have thought that this was close enough to Anaximander's qualityless substrate to make no difference, although Simplicius himself apparently acknowledges the distinction between them elsewhere (*in Phys.* 24,26ff.). On the other hand, Anaximenes' air is indeed infinite in extent – and Simplicius probably simply means to refer to this feature of his fundamental stuff; however, see n. 184 below.

184. i.e. this element underlying element would manifest itself as that into which things were destroyed – but no such element is observed (*Phys.* 3.5, 204b29-35).

185. *Phys.* 3.5, 205a7-b1.

186. This clause seems unwarranted by the argument, since the indifference principle appealed to in the following clause is supposed precisely to supply a reason why it should remain at rest (similar arguments are ascribed to Anaxagoras at *Phys.* 3.5, 205b1-13; the orginator of the indifference principle is Anaximander: *Cael.* 2.14, 295b11-16; cf. 295b16-296a21); so perhaps this clause should be secluded.

187. This argument is not to be found as such in *Phys.* 3.5; however cf. 205b1-24. esp. 12-13.

188. *Phys.* 3.5, 205a29-b1.

189. *Phys.* 3.5, 205b24-31.

190. i.e. disjunctively.

191. i.e. the argument proceeds by negated disjunction (*modus tollendo ponens*).

192. On uniformity, see 1.6, 221,8 above, and n. 126 ad loc.

193. The division is 'immediate', since given the antecedent it holds of necessity, independently of any further assumptions; i.e. the disjunction in question is an exhaustive one.

194. i.e. a *modus tollens* (cf. 1.6, 219,5-6 above, and n. 113 ad loc.); the argument is in fact of the form 'if *p* then either *q* or *r*; but neither *q* nor *r*; therefore not-*p*'.

195. *Cael.* 1.2, 268b14-16.

196. *Cael.* 1.2, 268b17-24.

197. *Cael.* 1.2, 269a3-4.

198. *Cael.* 1.2, 269a8-9; on the 'suppositions', see *in Cael.* 1.2, 12,6-11, and Hankinson, 2002, ad loc.; and see Hankinson, forthcoming 1, §IV, and n. 24 above.

199. 'Or five' seems to undercut the categorical statement of five lines earlier that there are only three such motions; but see *in Cael.* 1.6, 216,13-15 above, and n. 96 ad loc.; and 228,19-22 below.

200. i.e. what makes each body the body it is is its natural tendency to move; hence there will be as many natural bodies as there are such (actual) tendencies. The definition of nature as a principle of motion (and rest) is found at *Phys.* 2.1, 192b13-15; see further in general *Phys.* 2.1-2.

201. This follows of course only on the assumptions canvassed, namely that (a) form is determined by motile tendency, and (b) there are only a finite number of such tendencies. Atomists will deny (a) (see further below, 242,27,243,17: nn. 292-3). As regards (b), given that Aristotle (and Simplicius) allow that elemental form is not simply determined by the natural *direction* of motion (since in that case air would not be distinguishable from fire, nor water from earth), one might take issue with the claim that it has been established that there are only four sublunary elements: could there not be many more, each individuated from the others by its particular tendency to stratify in a particular region? After all, it is presumably supposed to be an empirical fact that there are only four sublunary elements. But then, it will be said on Aristotle's behalf given that the universe is finite (*Cael.* 1.5-6), there will not be room for an infinite number of them. But even that is not necessarily the case: we might posit an infinite Zenonian series of elements laid such that, given that the celestial radius is *r*, the first occupies the space out to a distance of 1/2*r*, the second the distance between 1/2*r* and 1/4*r*, the third the distance between 1/4*r* and 1/8*r* … etc. (see Hankinson and Matthen, forthcoming, ad loc.; and 1.6, 221,25: n. 129). There is of course nothing much to be said for such a view – but it does show that Aristotle cannot construct his theory of the elements of purely *a priori* grounds. But in fact, *pace* Simplicius, there no are reasons to suppose that he thought that he could: see Hankinson, forthcoming 1.

202. cf. *Phys.* 3.5, 204b4-22; the impossibility of the heavenly body being infinite has been argued in *Cael.* 1.5, that of the sublunary bodies being infinite in *Cael.* 1.6; see in particular *in Cael.* 1.6, 227,2-24.

203. The reference is to *Phys.* 3.5, 204b10-22 (n. 202 above); this betrays the extent to which Simplicius conceives of Aristotle's physical works (indeed his output as a whole) as a systematic whole (cf. *in Cael.* Pr., 2,29-5,38).

204. *Cael.* 1.6, 273a22-b15; cf. 219,1-220,25 above.

205. *Cael.* 1.5.

206. This phrase is awkward; Moraux (1965, 22), followed by Leggatt (1995, 72), deletes 'fire', yielding the sense 'it would be no less possible for the total sum to be infinite', which is attractive; but Simplicius' discussion (230,3-4) clearly shows that his text contained the word 'fire'; and so on balance it seems better to retain it. The sense is 'the totality of fire made up of all of the individual bits of it'.

207. sc. that the elements cannot be infinite in magnitude.

208. *Cael.* 1.5-6, esp. 273a7-21.

209. And this is impossible for any natural motion: cf. *Cael.* 1.6, 273a15-21; see also 217,16-218,10 above; and nn. 102-3 ad loc.

210. The reference Simplicius intends here is unclear; Heiberg cites *Phys.* 8.9, presumably referring to 265a17-20: 'it is not possible to travel along an infinite straight line, for there is no such infinite line; but even if there were, nothing could travel it, since what is impossible cannot come to be, and it is impossible to traverse an infinite line' (cf. 1.6, 218,2-3; n. 106). But that passage, although asserting the impossibility of infinite rectilinear motion, contains nothing directly bearing on the propositions here expressed, and certainly does not prove them. The 'first suppositions' here may well be the ones which drive arguments of *Cael.* 1.6, 273a7-21; at *in Cael.* 1.2, 12,6-11, Simplicius himself writes: 'with a view to establishing this [i.e. that there is a fifth heavenly element] on the basis of motions, he adopts these six hypotheses: (1) that there are two simple motions (circular and rectilinear); (2) that simple motion is of a simple body; (3) that the motion of a simple body is simple; (4) that there is <only> one natural motion for each body; (5) that for one thing there is one contrary; and (6) that the heavens move in a circle, as perception confirms'; on the nature and status of these 'hypotheses', see Hankinson, forthcoming 1; and n. 24 above.

211. cf. *Phys.* 8.9, 265a19-20; and see below, nn. 242, 250.

212. It is hard to see why Aristotle (and Simplicius) did not allow for the possibility of two (or more) unlimited amounts of stuff being arranged in serial parcels, like the squares on a three-dimensional chessboard (see Leggatt, 1995, 194, ad 274b18-22); a failure to deal with this vitiates the attempt to show that there cannot be more than one infinite body (see n. 225 below).

213. cf. fr. 59 B 1 DK: 'together were all things, infinite both in quantity and smallness'; in Anaxagoras' curious physics, everything is mixed through-and-through with everything else (59 B 4a, 6, 11 DK), excepting only the organizing Mind (59 B 12 DK); so Anaxagoras does indeed suppose that there are a variety of infinitely-extended substances. But since the point of the physics is precisely that they interpenetrate, it is hard to see how they can generate the alleged incoherences here, of there being a plurality of spatially-distinguishable substances.

214. i.e. the atomists, Leucippus and Democritus.

215. i.e. those who hold the worlds to be infinite in number.

216. The point of this obscure remark is unclear; Simplicius may mean simply that there are fewer conceptual problems standing in the way of the supposition that there is a single unified infinite body, since the (fallacious) considerations he rehearses here on Aristotle's behalf seem to make the idea of there being several distinct infinite bodies untenable; or this may be a reference to the physical model of Anaxagoras (above, n. 213), where the dissimilar substances are in a sense unified.

217. Reading *eiê* for *ara* of the MSS (cf. 230,13).

218. In other words, on the supposition that the infinite is made up of a limited number of types of things, then there must be an infinite quantity of at least some of the types.

219. i.e. the assumption of there being a finite number of types of stuff in the infinite, since it entails that there will be several distinct infinite quantities, is untenable.

220. Hence the only alternative is an infinite composed of an infinite number of types (the quantities of each of which could then be finite): but that too has been shown to be impossible (above, 228,12-24).

221. *Cael.* 1.6, 273b30-274a18; *in Cael.* 1.6, 222,4-225,34.

222. *Cael.* 1.5.

223. This supplement is required to complete the sense: I supply *oude hôs anomoiomeres* after *einai to apeiron* at 231,4.

224. Thus not every motion in a body which is induced by something else will be unnatural and forcible (in this strong sense) for that body: something is moved unnaturally (or counter-naturally) in this sense only when that movement actively militates against the body's natural motile tendencies.

225. Once again, this contention relies on the (false) supposition that any infinitely-extended body must exhaust the whole of any infinite space: 229,28-30, and n. 212 above; cf. *Phys.* 3.5, 204a20-6; 204b28-9; 205a23-5; and see *in Cael.* 1.5, 207,12-16: n. 43.

226. i.e. its motion does not require that there be space empty of it into which it can move.

227. Aristotle's preferred definition: n. 228 below.

228. Aristotle defines the place of a body as the innermost limit of the body which surrounds it (*Phys.* 4.2, 209a31-b6; 4.4, 211a23-212a14), and he explicitly rejects the option canvassed by Simplicius here (*Phys.* 4.4, 211b9-29); however, Simplicius may simply be pointing out that, on the impossible (on Aristotelian grounds) supposition that an infinite mass existed, it could be said to have a place only in some such manner. Aristotle himself acknowledges difficulties created by his definition for the idea of the place of the universe as a whole (since nothing – not even nothing itself – lies outside it) at *Phys.* 4.5, 212a31-b22. Simplicius himself, however, is receptive to the idea of place as pure extension: see Sorabji, 1988, chs. 11-12, esp. 206-11.

229. The idea that the capacity to act or be affected was definitional of body became canonical in later Greek philosophy under the influence of the Stoics (and to a lesser extent the Epicureans): Sextus *PH* 3.38. It was never accepted in its full generality by Platonists, who wanted to allow for non-physical causation – but that is irrelevant to the matter at hand here.

230. This is an expansion of Aristotle's statement in the lemma (*Cael.* 1.7, 275a7-14); but the axioms are established (insofar as they are) at *Cael.* 1.6, 273b30-274a3; and see also *Phys.* 7.5, 249b29-250a16; see further 232,19-20 below.

231. i.e. there is no such thing as instantaneous motion: see *Cael.* 1.6, 274a3-13; *in Cael.* 1.6, 223,10-20, and nn. 144-5; and see also *Phys.* 4.8, 215a24-b22; 6.1, 231b18-232a18; 6.10, 240b30-241a6.

232. i.e. something finite. The clarity of the argument would have been helped had this quantity also been assigned a letter, a fact which Simplicius notes elsewhere: 235,8-9 below. Let it be F (see Fig. 1). It also helps to assign a letter to the effect produced, which is *ex hypothesi* the same in each case: this may be G. Then the argument runs as follows: if the finite B produces a certain change G in the infinite A in time C, then, some smaller finite agent than B, call it D, will effect the same change G in something smaller (and also finite), namely E. But as D stands to B, so E stands to some (finite) patient F; and hence B should produce G in F in the same manner in C as it affects A; but A is infinite and F is finite. It is a feature of this argument that the lesser mover does not move (or in general affect) the same body to a lesser degree, but rather causes the same change in a lesser body – this is a new *démarche*: but the same principles of proportionality apply. It might have been better to translate the verbs of motion – *kinein, kineisthai* – by something more general – cause, change, effect, or the like, since as Aristotle makes clear in the lemma, the ambit of the discussion is broader than simply that of locomotion and includes change in any category (274b33-275a10): but I have opted for consistency in translation.

233. These arguments are structurally isomorphic with those of *Cael.* 1.6, 274a3-19; see *in Cael.* 1,6, 222,33-225,34.

234. The 'second indemonstrable' is one of the five fundamental Stoic argument-schemata, corresponding to the modern *modus tollens*: if *p* then *q*; not-*q*; therefore not-*p*; this is the same pattern Simplicius referred to in earlier contexts as 'denying the consequent' (219,5-6; 227,33). I have slightly altered Heiberg's punctuation here to restore the sense.

235. See n. 230 above.

236. The clause in square brackets is clearly redundant, and could easily be an intrusion; on the other hand, such prolixity is not untypical of Simplicius.

237. i.e. Given that F:BF::D:E (and that D:E::F:BF: see n. 236 above), then it follows that D:F::E:BF.

238. i.e. at 275a15, 17, Aristotle calls the finite moved object B, from which F is subtracted as a submagnitude: but then at 275b17, 19, he calls the same object BF, as though the whole were the *sum* of B and F.

239. i.e. and not instantaneously, which is impossible: *Phys.* 4.8, 215a24-b22; 6.1, 231b18-232a18; 6.10, 240b30-241a6; and cf. *in Cael.* 1.6, 223,18-20: nn. 144-5.

240. 231,22-233,10.

241. This option is added for the sake of completeness, not to make the disjunction of options exhaustive; cf. nn. 190, 191, 193 above; strictly speaking, the further possibility of the infinite's being moved by something finite in an infinite time should also be canvassed here.

242. In other words, generation must be a completable process: 229,17-21, and nn. 210-11 above; cf. 1.6, 218,2-3: n. 106.

243. Above, 233,28-33: n. 230.

244. Infinite A brings about a certain change in infinite B in finite time CD. Now consider a finite proper part of B, namely E: A will effect the same change in E as it does in B in a part of CD, namely D.

245. cf. n. 232 above. Since the ratio of D:CD is rational, and E is finite, there must be some finite magnitude F which is such that D:CD::E:F.

246. D:CD::E:F; hence D:E::CD:F.

247. Since dynamics are proportional, given that A affects E in D, A will affect F in CD; but F, being finite, must be less than B, which is also supposed to be affected by A in the same time CD: which is impossible.

248. sc. the proper part of it E.

249. A close paraphrase of *Cael.* 1.7, 275a32-b2.

250. Since it can never arrive at an end, it makes no sense (Simplicius thinks) to suppose that it approaches one: cf. n. 242 above.

251. 234,16-17.

252. The syllogism is in the second figure mood Camestres ('A belongs to all B; A belongs to no C; so B belongs to no C': AaB; AeC $\vdash$ BeC); the converse of the conclusion can be derived, as Simplicius says, simply by applying the general rule that universal negatives convert (AeB $\dashv\vdash$ BeA); however, by simply inverting the order of the premisses, the converse conclusion can also be validly inferred in the mood Cesare ('A belongs to no B; A belongs to all C; so B belongs to no C': AeB; AeC $\vdash$ BeC).

253. See *Phys.* 3.6-7.

254. The last clause is omitted from the lemma; but this is clearly an oversight.

255. This issue is taken up in *Cael.* 1.8-9.

256. For an explication of this terminology, see 238,5-15 below.

257. See n. 213.

258. The inference thus presented is in the first-figure mood Celarent (A belongs to no B, B belongs to all C, so A belongs to no C: AeB; BaC $\vdash$ AeC), a point which Simplicius makes more of below.

259. Another syllogism in Celarent: above, n. 257.

260. From which the omitted conclusion indeed follows, by way of Cesare (on which see n. 252 above): see n. 263 below, on the omitted intermediate conclusion.

261. The first syllogism is in the mood Camestres: above, n. 252.

262. The second syllogism, using the conclusion of the first as a premiss, is also in Camestres; the whole argument goes as follows: PaS, PeI $\vdash$ SeI; SaO, SeO $\vdash$ OeI ('P' = perceptible; 'S' = spatially-located; 'I' = infinite; 'O' = outside the heaven).

263. Simplicius here alludes to the Aristotelian mechanism of reducing syllogisms in the second and third figures to the 'perfect' first-figure syllogisms by way of certain conversion rules (in this case crucially the conversion of universal negatives, noted above: n. 252): *An. Pr.* 1.7, 29b1-26.

264. The Stoic logicians originally formulated four *themata*, or ground-rules for inferential reduction, in their hypothetical syllogistic, of which only the first and the third are securely known. The first specifies that, from the negation of the conclusion in a valid (two-premiss) argument plus one of the premisses, the negation of the other premiss can be validly inferred (Apuleius *On Interpretation* 191,5-10 = 36I Long/Sedley); the third is given here (and also at Alexander *in An. Pr.* 278,11-14 = 36J Long/Sedley); the precise structure of the second and fourth *themata* is not preserved, but Sextus tells us that all that the second, third, and fourth *themata* could achieve was summarised in the 'logical theorem for the analysis of arguments': 'when we have the premisses from which some conclusion is deducible, we potentially have that conclusion in the premisses' (Sextus *Against the Professors* 8.231). The third *thema* amounts to the following: if arguments P, Q $\vdash$ R, and R, S $\vdash$ T are valid, then P, Q, S $\vdash$ T is also valid. In this case, the R is the conclusion of the intermediate syllogism, namely that no spatially-located body is infinite: above, 236,19-22, and n. 259.

265. The first syllogism (in Celarent: above, n. 257) reads: every spatially-located body is perceptible, no perceptible body is infinite, so no spatially-located body is infinite; the second (also in Celarent): no spatially-located body is infinite, what is outside the heaven is spatially-located, so there is no infinite body outside the heaven. The complete inference may be symbolized: IeP, PaS $\vdash$ IeS; IeS, SaO $\vdash$ IeO (see n. 262, for the interpretation of the letters).

266. *Cael.* 1.8-9.

267. Simplicius is probably thinking of *Cael.* 1.9, 278b8-279a11; but all that

Aristotle directly establishes there is that there can be no body outside the confines of the universe; he does not specifically prove that there can be no finite body outside it (although that evidently follows from the general conclusion).

268. cf. *Cael.* 1.9, 279a11-279b3.

269. *Cael.* 1.7, 275b6-7.

270. *Cael.* 1.2.

271. Above, 236,10.

272. This seems to be the official version; but note that at 236,10 above, Simplicius describes certain arguments as both more concrete and more general.

273. *endoxa*, characterized at the beginning of the *Topics* as 'those which commend themselves to the great majority of people, or to the wise, that is either to all of the wise, or the great majority of them, or to the most knowledgeable and reputable' (*Top.* 1.1, 100b21-3)

274. See the opening sentence of the treatise: 'the purpose of the treatise is to discover the method by which we will be able to draw conclusions about any problem set for us on the basis of reputable opinions' (*Top.* 1.1, 100a18-20); the result is 'dialectical argumentation', which proceeds from *endoxa* (100a30-b1).

275. Literally 'places', *topoi.*

276. The inference is once again in the mood Cesare: above, n. 252.

277. *Cael.* 1.5.

278. *Cael.* 1.6, 273a22-b15; 219,3-220,25.

279. i.e. the type of demonstration found at *Cael.* 1.6, 273b30-274a3; 1.7, 274b33-275a14.

280. Which are by nature uniform (in substance: the issue here does not have to do with uniformity of density: see 1.6, 220,5-7: n. 121; 221,7-25: nn. 127-9).

281. See *Cael.* 1.2, 268b26-269a2; *in Cael.* 1.2, 17,18-33; the point is that any natural movement undertaken by a composite body will in fact be caused by the simple bodies of which it is composed; and hence the argument concerning simple bodies can be generalized to cover any natural motion at all.

282. This is the crucial, mistaken assumption which drives so much of Aristotle's argument about the infinite, namely that if something is spatially infinite, it must exhaust all the available space: cf. n. 225 above.

283. *Cael.* 1.7, 274b29-32.

284. cf. 238,21-9 above.

285. i.e. in the earlier passage: n. 283.

286. i.e. the considerations adduced in regard to it are purely geometrical ones, as opposed to the physical characteristics (heaviness, lightness, their tendencies towards their respective places) which are invoked in the case of the rectilinearly-moving bodies.

287. 239,9-19.

288. *Phys.* 7.1; cf. 8.4.

289. *Phys.* 8.10, 266a23-b27, referred to by Aristotle in the lemma above as *On Motion*; for this designation, see *Cael.* 1.5, 272a30-1; see also *in Cael.* 1.6, 226,19-21, and n. 268.

290. i.e. in the passage of *Phys.* 8.10 cited above (n. 289); cf. *Cael.* 1.6, 273a22-b30, for the argument that an infinite object will be either infinitely light or infinitely heavy (*in Cael.* 1.6, 219,1-222,5).

291. The parenthesis is correct: the thesis that every mobile body is moved by a distinct mover is irrelevant to the point at issue (since that mover may be intrinsic to the body in question and also incorporeal); and while Aristotle does adopt the thesis mentioned in *Phys.* 8.4, 254b7-256a3 (and 7.1, 241b24-242a16), there is little hint of it to be found in *Cael.* (but cf 1.7, 275b25-9 below; and 241,5-18).

292. i.e. by force.

293. Some such phrase needs to be understood here to make sense of Simplicius' argument; but the phraseology is odd, and more radical surgery may perhaps be needed (the sense is clear enough however).

294. Reading *kinountos* after *tinos*; the claim is not that is invariably so moved, but that rather that it can be, and since it can be, it will be (for this modal inference, see *Cael.* 1.12; and Hankinson and Matthen, forthcoming, ad loc.).

295. Presumably simply in the sense of 'more general': i.e. it does not rely on any of the particular principles of proportion which underpin the earlier arguments.

296. 239,12-19; 239,36-240,30.

297. *Cael.* 1.6, 273b30-274a18; *in Cael.* 1.6, 222,4-225,34; cf. 230,21-8 above.

298. Heiberg takes the reference to be to *Phys.* 2.1, 192b8-b10; but there it is a thing's *internal* nature which is said to be responsible for its natural tendency to move. More probably Alexander intends to recall *Phys.* 8.4, where Aristotle seeks to demonstrate that everything is moved by something distinct from itself (254b12-33), and that even in the case of the natural motions of the elements, something external to the elemental mass needs to act before that mass actualizes its potential to move (255b13-31).

299. *Cael.* 1.7, 275b18, above; on Alexander's view, Aristotle does not deal with things which move as a result of their own internal tendencies here; Simplicius (rightly, given the context of the argument) thinks this unlikely.

300. The overall structure of the argument is as follows: if there is an infinite mover then either (a) it is moved by something else, or (b) it is self-moved, or (c) it moves naturally; but not (b), since what is self-moved is an animal, and there is no infinite animal; and not (a), since its external mover would have to be infinite too, and there cannot be two infinites; but not (c) either, since what moves naturally will also at some time be moved unnaturally, and hence by some other infinite mover, and there cannot be two infinites; hence there is no infinite mover.

301. Reading *diôrismenon* (the reading Simplicius adopts: 245,31-2) in place of the MSS *diôrismena* here; the latter yields the sense 'if the totality is not continuous, but is formed of things divided by the void': the sense is not affected.

302. For the infinity of atoms in Leucippus, see Simplicius *in Phys.* 1.2, 28,4-15 (= 67 A 8 DK); for Democritus, see also the fragment from Aristotle's *On Democritus*, reported by Simplicius at *in Cael.* 1.10, 294,30-295,22 (= 67 A 37 DK). Democritus held that not only was there an infinite number of individual atoms, there was also an infinite number of atomic types: *in Phys.* 1.2, 28,8-10 (= 68 A 38 DK).

303. For Epicurus (who denied the infinity of atomic types) see *ad Hdt.* 41-3.

304. i.e. the atoms.

305. A view also endorsed by the renegade Peripatetic Strato: fr. 65a Wehrli.

306. Strictly speaking, these last two properties are distinguishing features of *groups* of atoms, or of atoms considered in relation to one another: cf. Aristotle *Metaph.* 1.4, 985b4-20.

307. cf. *in Cael.* 1.10, 294,30-295,22 (= 67 A 37 DK); *in Phys.* 1.2, 28.15-27.

308. This is same phrase (*kata tên phusin*) which is standardly translated as 'natural'; I have preferred the alternative here in order to avoid suggesting that the motions of the atomists' bodies are natural in anything like the Aristotelian sense.

309. See in general *Cael.* 3-4; especially 3.2, 301a20-b16; 4.2 308b28-309b28. The atomists of course have answers to these objections: things which apparently rise are only *relatively* light – they are less dense – and tend to settle out above the heavier stuffs by extrusion: see *Cael.* 1.8, 277a27-b9; *in Cael.* 1.8, 267,29-268,17.

310. See *Cael.* 1.2, 268b26-269a2; 269a4-5; 269a28-30.

311. Once again, this argument depends upon accepting – as of course the atomists would not – a particular conception of weight and lightness as being tendencies oriented towards particular determinate places. It is not clear whether Democritus held that atoms have weight. Aristotle (*Gen. Corr.* 1.8, 326a9-10 = 68 A 60 DK), Theophrastus (*Sens.* 61 = 68 A 135 DK), and Simplicius (*in Cael.* 4.4, 712,27-9, 68 A 61 DK) all hold that he did; Aëtius (1.3.18, 68 A 47 DK) says that it was an amendment of Epicurus (who certainly did think so: *ad Hdt.* 10.54; Lucretius 1.358-67; 2.184-215). Epicurus thinks that this amounts to no more than the tendency to move in a certain direction (the way everything moves – downwards); that direction need not itself be determined by structural facts of the universe in Aristotle's manner (but see *in Cael.* 1.8, 268,1 below).

312. At *Phys.* 4.8, Aristotle offers a series of arguments against the possibility of motion in the void: (1) at 214b12-28, he argues that void cannot be responsible for anything, including natural motions; (2) at 214b28-215a24 he contends that void will be inimical to motion since it admits of no internal distinctions; (3) at 215a24-216a23 he offers dynamical considerations against its postulation: speed of motion is a function of the mass of the mobile object and the inverse of the resistance of the medium through which it passes, but there is no resistance in a void, and hence the motion should be infinitely fast, which is impossible (cf. nn. 132, 143-4, 153 above); finally (4) at 216a23-b21 he argues that the void is incoherent conceptually and can serve no useful function in physical argument. None of these considerations precisely mirror the considerations advanced here, although (2) makes use of some of the same material (esp. 215a6-11).

313. This follows only on the assumption that 'natural' and 'unnatural' jointly exhaust all possibilities for motion; but in turn that is reasonable only if natural motions include the self-movement of animals – but those are not determined by natural places.

314. i.e. that of 242,27-243,10 above; for this principle applied to motions, cf. *Cael.* 1.2, 269a32-269b2.

315. *epagôgê*: not here a species of inference, but rather confirmation by enumeration of cases.

316. i.e. the passage introduced by 'furthermore' (276a12-16) is a separate argument for the same conclusion, not a part of the original argument (which is thus supposed to stand on its own).

317. i.e. omitting the inferential particle *dê*: I have punctuated more heavily than Heiberg after 'etc'. to make the point clearer (namely that Aristotle would have separated these sentences logically and syntactically had he been doing what Alexander supposes him to have been doing). One of Aristotle's MSS (E) does omit the *dê*; and Solmsen (1960, 303, n. 49) conjectured *de* in its place, which has the same effect.

318. i.e. if there were only one natural direction of movement (as favoured by Epicurus: n. 120 above), then every item would have the same natural tendency; and insofar as there could be unnatural motion (i.e. motion against the general tendency of everything to move 'downwards') it could not be the natural motion for anything else. But here Aristotle is prone to the charge of begging the question in favour of his own conception of the nature of natural and unnatural motion: Epicurus can simply deny that what is unnatural for one is natural for another. But Aristotle can be defended: he does claim that it is an empirical fact that there are contrarieties of natural place (e.g. fire has to be forced downwards, while earth moves naturally that way); but again, if the contention is construed as an empirical hypothesis, Epicurus has a response available in the extrusion theory: n. 309 above. See further Hankinson, forthcoming 3.

319. Which was the atomist view: atoms differed according to 'shape, arrangement, and position' (Aristotle *Metaph.* 1.4, 985b13-19 = 67 A 6 DK), and in no other way.

320. *Cael.* 1.7, 274b18-19.

321. Heiberg does not mark the end of the quotation from Alexander: but it must fall here.

322. See above, 236,10.

323. i.e. there is no separate argument against the infinity of the atoms here (as Alexander thought: 245,7-17); rather the possibility of there being an infinite world of atoms and the void falls if there cannot be a world of atoms and the void (on the atomists' lines) at all.

324. cf. 1.6, 274a24-8.

325. *Phys.* 3.4-8.

326. In the Proem to *in Cael.*, Simplicius discusses the meaning of the title of the work, and its *skopos*, or proposed subject-matter (1,2-6,27), a vexed question among the ancient exegetes, and concludes that properly-speaking its subject is the elements, including the sublunary ones (5,23-34), since they have the form they do as a result of the configuration of the heavens (cf. *Cael.* 1.2, 268b13-14; and see Matthen and Hankinson, 1993).

327. Alexander here paraphrases, without quoting, Aristotle's remarks at 276a19-22; Simplicius quotes exactly at 247,5-7.

328. Reading *tauta* for *touto*.

329. The sense would be improved if the text read 'of some world' rather than 'of the world' (*tên diathesin kosmou tinos* instead of *tên tou kosmou diathesin*); the point is, as Simplicius goes on to say, that things 'with no definite position' are those which are not organized into any cosmic structure, and hence, for the purposes of this argument, not any body whatsoever, organized or not, outside this cosmos, as the text here would suggest. But I hesitate to emend the text here.

330. *Cael.* 1.6-7.

331. 1.7, 275b8-9.

332. 1.7, 275b9.

333. i.e. categorically.

334. cf. *Cael.* 1.2, 269a2-9.

335. Because there is no contrary to circular motion: *Cael.* 1.4; Simplicius should strictly speaking have said that what moves in a circle cannot be moved unnaturally, and cannot rest in any way at all.

336. i.e. of one another: lateral movement to the left is not contrary to lateral movement to the right, on Alexander's view; rather they are together generically contrary to natural motion downwards; see n. 337 below.

337. This principle is first enunciated at *Cael.* 1.2, 269a10; cf. *in Cael* 1.2, 12,11; 19,17-29, etc.; it was challenged by the first century BC Peripatetic Xenarchus (*in Cael.* 1.2, 55,25-56,25), on the grounds that each of two opposing vices may be considered to be contraries of the same virtue: see Hankinson, forthcoming 3.

338. *Cael.* 1.2, 268b11-269a19.

339. cf. *Cael.* 1.2, 269a9-18.

340. i.e. *Phys.* 5.1; cf. 1.5, 188a33 ff.; Simplicius supposes that the *Physics* formed the first part of a long, unified treatise on natural science, including *de Caelo, Meteorologica, de Generatione et Corruptione, de Anima*, the *Parva Naturalia*, and the biological texts (*in Cael.* Prologue, 2,29-3,12).

341. i.e. even what is intermediate between being *F* and not being *F*, and which is in a state of becoming not-*F*, is already in a sense contrary to *F*, even though it is not, in a strong sense, the polar opposite of *F* (cf. *Phys.* 5.1). Aristotle does indeed allow that the notion of contrariety is an elastic one, and will on occasion tolerate looser usages: cf. *Metaph.* 5.10, 1018a26-31: see Hankinson, forthcoming 3, §II.

342. Simplicius makes the point elsewhere: *in Cael.* 1.2, 19,20-4; cf. 51,23-8.

343. See *Cael.* 1.2, 269a9-10; 269a32-5; 1.3, 269b35-270a3; 3.2, 300a20-7.

344. 276a23-4.

345. i.e. 'upwards' with respect to this cosmos: that is the sense of 'from here to there'.

346. Reading *touto* in place of *to*; again 'upwards' with respect to this cosmos.

347. 276a30-b18 below.

348. 276b14-15.

349. 276a30-b18.

350. 276a30-b7.

351. Since its nature is to move towards the centre of that world, which is upwards in relation to the centre of this one.

352. i.e. even though this is one and the same motion (away from here and towards there), none the less it will exhibit contrariety, since the centre here and the centre there are identical in form (249,20-4), and so it will possess the opposed properties of moving both towards and away from a centre.

353. Some such supplement is required for the sense, as the subsequent lines (250,7-9) show. Perhaps the text should read *eis touto kata phusin estin hê phora autêi, vel sim.*, in place of *touto kata phusin estin autêi*: 'movement towards this place [i.e. the one where it rests naturally] is natural for it'.

354. 276a23-6.

355. i.e. it is the fact that there is only one natural motion for each body which, in conjunction with the hypotheses canvassed, yields the contradiction; but it would do so equally well on Alexander's interpretation (above, 249,3-15).

356. At 252,23 Simplicius quotes Aristotle as having written *katô* here; the MSS of *Cael.* all read *epi to meson*, 'towards the middle': the sense is unaffected.

357. sc. naturally: there is of course nothing absurd about earth moving upwards and fire downwards unnaturally.

358. The fundamental qualities of fire are heat and dryness, those of earth coldness and wetness (*Gen. Corr.* 2.3; cf. *Metaph.* 4.1); of these, heat and cold are the primary or active qualities (*Gen. Corr.* 2.2, 329b16-33) which is presumably why Simplicius singles them out here. Moreover, 'heavy and light are neither active nor passive: things are not called heavy and light because they affect, or are affected by, other things' (*Gen. Corr.* 329b20-2). But while they are from the perspective of *Gen. Corr.* relatively unimportant properties, they are central to the arguments of *Cael.*

359. i.e. the intermediate elements, air and water.

360. Simplicius' language here is ambiguous: 'the middle' could could refer to (a) the middle of the other world (as is perhaps more natural); but equally it might mean (b) 'the middle <of this world>'. The argument could be constructed either way (if it is taken in the sense of (a), then the method will be to generate a contradiction: the earth of the other world will have a tendency both to move towards the centre there and towards the centre here); but Aristotle's own argument (276b11-14) clearly suggests that we should take it in sense (b), and that seems the likelier interpretation for Simplicius as well (see 251,30-36 below).

361. *Cael.* 1.2, 268b26-269a12; cf. *in Cael.* 1.2, 12,6-11.

362. If interpreted according to possibility (a) of n. 360, this will have to be read as meaning that each separate world determines its own simple motions in accordance with its own spatial geometry; but more probably Simplicius intends to draw the stronger (albeit unwarranted and question-begging: see n. 363 below) conclusion that there can be only one centre and one periphery to supply the orientation for a unique set of natural movements. On these issues, see Hankinson and Matthen, ad loc.

363. Again this paragraph is susceptible of interpretation either according to sense (a) or according to sense (b): n. 360 above. But again it is likely that Simplicius is aiming for the stronger conclusion: since earth is earth no matter where it is, and since what something is determines (and is partially determined by) its natural motile tendency, then earth here and earth in any other world will possess exactly the same motile tendencies – but then they will each seek the same places. The argument works of course only if sameness of motile tendency is determined absolutely with respect to a single determinate point or region of space, which is to beg the question: nothing seems to rule out the possibility of any and every piece of earth being alike in the sense that they will all seek the centre of whatever cosmos they happen to find themselves in.

364. For rather clearer expressions of the principle that only something which can have come to be (i.e. whose process of generation can be completed) can really be said at any time to be coming to be at all, see *Cael.* 1.7, 274b13-15; *in Cael.* 1.7, 229,17-25: n. 211; 234,15-17: n. 242; 235,31-3: n. 250; and cf. *in Cael.* 1.6, 218,2-3: n. 106; and n. 418 below.

365. Circular motion is an *energeia*, an activity, something which is complete at every moment of its existence (cf. *in Cael.* 1.2, 65,22-8), rather than a *kinêsis*, a process. At *Phys.* 3.1, Aristotle defines *kinêsis* as 'the actualization of what is in potentiality, insofar as it is such' (cf. *Cael.* 4.3, 310a20-b1). Rectilinear motion, since it is motion *towards* something, is a *kinêsis*, and thus irreducibly involves the notion of potentiality. Circular motion, on the other hand, has no beginning and no end (cf. *Phys.* 8.8, 261b27-265a12). Strictly speaking, though, eternal circular motion does not involve coming to be anything at all.

366. cf. *Cael.* 1.7, 276a2-4; *in Cael.* 1.7, 243,3-5.

367. i.e. at the centre and the extremity of *this* world.

368. i.e. in each of the postulated worlds.

369. i.e. if the original hypothesis entails an impossibility, it must itself be impossible: see further below, 253,18-21. For this fundamental modal principle, see *An. Pr.* 1.23, 41a22-32.

370. It is not clear why Alexander thinks that worlds must be arranged so as to be in contact with one another – it is presumably enough for them to be distinct and spatially separated. Alexander may be relying on the general Peripatetic belief that there can be no absolutely empty space (cf. *Phys.* 4.6-9); but even if the worlds touch one another, given that they are spherical there will still be empty interstices between them.

371. This supplement obviously needs to be understood here.

372. See n. 370 above.

373. See 276a14-15: n. 356 above.

374. i.e. when it has crossed the boundary between the worlds and is making its way down through this world's layers of ether.

375.

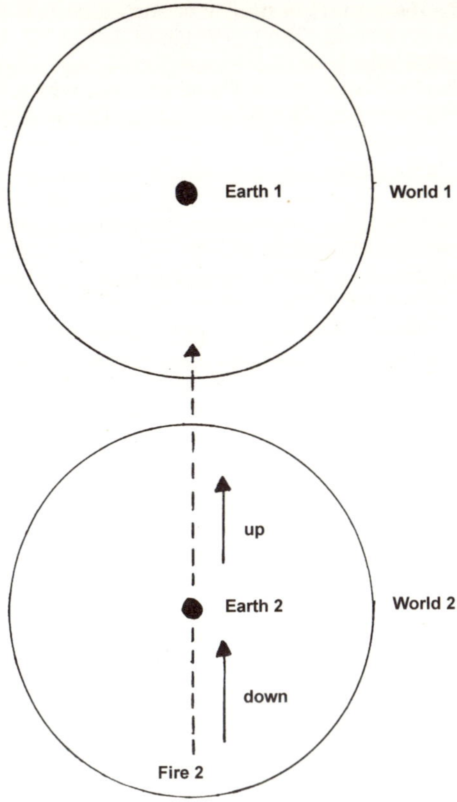

376. i.e. there.

377. i.e. the fire in the other world.

378. Although what Simplicius says here is technically correct, it seems unduly harsh on Alexander, who probably meant 'above' and 'below' merely metaphorically.

379. cf. *Cael.* 1.2, 269a9-18; 1.3, 269b35-270a3; n. 345 above.

380. *Cael.* 1.2, 269a8-9.

381. Heiberg ends the quotation from Alexander here, and marks the lines from 'So if, on the assumption' to 'and away from this middle' as being Simplicius' own comments. It seems more likely that they continue Alexander's remarks.

382. See 252,5-9; n. 369 above.

383. i.e. as long as they are all called worlds in the same sense. But neither Aristotle nor Simplicius apparently consider the possibility of there being another world in which there is an earth-like substance (which only tends towards the centre there) and a fire-like substance (which only tends towards that periphery), and likewise with the twin-world intermediate elements. No such supposition has been ruled out by any of the arguments thus far – and nor will it be. It is a further question whether such a world really ought to be called 'a world' (one might want to claim that it was specifically distinct from this one, but surely it is generically similar: and 'world' may perfectly well be a generic term); but in any event, resolution of that terminological issue will have no bearing on the substantial possibility of such a supposition. On homonymy in general, see *Cat.* 1, 1a1-12; *Top.* 1.15, 106a1-107b35; 2.3, 110a23-111a7; *Soph. El.* 17-24, 175a31-180a22.

384. Reading '*ta*' for the '*to*' in Heiberg's text, which would have Simplicius imply that earth, water, *and* air have a natural downwards tendency.

385. i.e. each of the elements.

386. See n. 369 above.

387. Again this follows only on the assumption that determining natural motion is itself determined by reference to a particular spatial location, which is to beg the question: above, nn. 362-3.

388. For the principle that the closer things are to their natural places the faster they move, see 1.8, 277a27-33 below; and 263,13-267,6.

389. i.e. the mere fact of being very far removed from the natural point of orientation cannot be enough to induce an actual change of nature for the body in question (as it would have to do if the body lost all inclination to move towards that point). But Simplicius does not allow for the possibility that earth in this world might be affected by a tendency to move towards the centre of some other world, but that that tendency would be negligible in comparison with its tendency to move to towards the (much closer) centre here – and similarly for all other worlds (this would make the movements – of heavy objects at least – very roughly analogous to their behaviour in a Newtonian universe). No doubt Simplicius would still reply that that would be to give them more than one natural motion; and also perhaps that the weak nisus towards the centre of other worlds would be (since it could never cash out in actual motion) pointless – but nature never does anything which is pointless (cf. *Cael.* 1.4, 271a22-33).

390. *Cael.* 1.8, 276a22-6.

391. *Cael.* 1.2.

392. *Cael.* 1.2, 268b14-26.

393. *Cael.* 1.2, 269a2-b10.

394. *Cael.* 3.2-5.

395. This punning is at the expense of Democritus: he is said to have written a *Great Diacosmos* or world-ordering (although this was also attributed to Leucippus: Diogenes Laertius 9.39-40, 46), as well as a *Little Diacosmos* and a *Cosmography* (*ibid.* 41, 46); according to him, worlds were generated by the action of vortices on atoms moving in a random manner: Simplicius *in Phys.* 327,23-6.

396. Reading *hôs* for the *hoti* of the MSS.

397. The text here is hardly satisfactory, and my translation follows a somewhat strained construal of the Greek – but I can make no sense of the more natural rendering: 'and that the natural bodies of the mathematical objects which are unmoved are most particularly known through their motion.'

398. *Cael.* 1.2, 269a9-18 (*in Cael.* 1.2, 19,11-20,10); 1.7, 274b29-32 (*in Cael.* 1.7, 231,3-19). The argument seems to be the following: if the bodies move forcibly (but not naturally) they will do so in any and every direction (which is of course an assumption of Democritean physics); but then they will move forcibly in contrary directions; but if they move forcibly in one direction, they will move naturally in the contrary direction and not forcibly, and so the asusmption that they will move forcibly in all directions is contradicted. Once again, this argument begs all the relevant questions, in this case those involving the notions of contrariety and those of natural and unnatural motions.

399. The text is awkward here, and some text may be missing; but this seems to be the sense.

400. The argument apparently is that if the body 'is of such a kind as to move', then, since it *is* of such a kind, it will have a natural motion. But this is plainly inadequate: it may be of such a kind as to be moveable without that entailing that it must have some determinate tendency to move itself.

401. Again this contention, that the unnatural presupposes the existence of the natural, is one no atomist need accept: rather, all motions may be unnatural in the sense that they are not caused by any particular internal tendency of the moving object to move; equally, they may all be natural in the sense that every body is of a nature to allow the imposition of motion upon it.

402. Aristotle's own definition of place: see *Phys.* 4.2, 209a31-b1; 4.4, 211b6-14, 212a6-7.

403. i.e. they serve to mark off the directions in question, but they do not define them in the sense of giving their essences.

404. By 'intermediate elements' here Simplicius clearly means the sublunary four, even though such expressions elsewhere refer only to air and water; the usage is perfectly understandable in context, since the centre and the extremity define limits within which the elements exist and express their natural tendencies.

405. In Aristotle's technical vocabulary a premiss is immediate (*amesos*) if there is no explanatory middle term which can relate the subject to the predicate in the form of a valid syllogism: thus 'all men are animals' is immediate (it is a definitional truth of humankind), while 'all men are mortal' is not, since men are mortal in virtue of the fact that they are animals, a fact which can be expressed syllogistically: all men are animals; all animals are mortal; so all men are mortal. See *An. Post.* 1.3, 72b18-24; 1.9, 75b37-40; 1.33, 88b33-89a4; 2.9, 93b21-8; etc.

406. i.e. the supposition is that the fact that the hypothetical alternative world is also surrounded by a spherical heaven makes the directions there identical in type to those here.

407. i.e. the hypothetical elements in another world.

408. *Cael.* 1.2, 269a8-9.

409. 276b29-32.

410. See n. 326 above.

411. Reading '*autôn*' in place of Heiberg's '*autou*'.

412. i.e. by the processes of mastication and digestion.

413. Aristotle distinguishes the nutritive (or vegetative) soul, by which the living thing derives and metabolises nourishments, from the other capacities of living things. It is the most widespread such capacity, being common to plants as well as animals: *De An.* 2.2, 413a25-b9; 2.4, 414a29-b1.

414. *analusis* in geometry is the procedure of moving towards axioms or first principles from evident facts which are consequences of them; the procedure is exemplified in famous passages in *Meno* (86C-89A: the geometrical model of hypothesis) and in particular in the *Republic* (509D-511E: the divided line; esp. 510B-D). Descartes invokes the distinction too: the *Meditations* is analytic in structure, explaining how one arrives at the fundamental principles; these may be followed (as here) by a synthesis, which consists in moving in the other direction and deducing the theorematic consequences of those principles (as exemplified in *Principles of Philosophy*); and the same distinction is famously to be found in Kant. The methodology – both in specifically geometrical contexts, and in more general philosophical ones – raises a number of pressing epistemological and logical questions which cannot here be addressed. In Plato, the notions is allied to the so-called 'method of hypothesis' (cf. also *Phd.* 99D-107B, esp. 99E-102A); the best discussions of both issues are still to be found in R.

Robinson, *Plato's Earlier Dialectic* (Oxford, 1953), chs 7-11), and 'Analysis in Greek geometry', in Robinson, 1969.

415. *Phys.* 1.5-7, esp. 188a30-b26; cf. 5.1-2.

416. *Phys.* 5.1, 224b30-1 (cf. 5.5); see n. 341 above.

417. cf. *Metaph.* 5.10, 1018a28-9: however, contrary to Simplicius' practice here, Aristotle treats this as only one possible definition of 'contrary' (*enantion*), which itself forms a sub-class of opposites in general (*antikeimena*); cf. *Top.* 4.3.

418. *Cael.* 1.7, 274b10-22; *in Cael.* 1.7, 229,17-25: nn. 210-11; see also n. 364 above.

419. *Cael.* 1.4.

420. cf. *Cael.* 1.4, 271a21-3.

421. This is a somewhat curious contention. Consider a fixed point p_1 on the orbit of a revolving body, and a mobile point m on that body. If m starts from p_1, then describes half an orbit and arrives at p_2 diametrically opposed to p_1, it can get no further away from it. But the point is that there is no natural stopping-place for m there, and no reason for its motion not to continue in such a way as to bring it back to p_1 again.

422. i.e. p_2: see n. 421 above.

423. i.e. 'moving along a (particular) semi-circle' is not an intrinsic characterisation of the motion of any revolving body – rather it simply happens to complete indefinitely many traverses of arbitrarily-determined semi-circles as a consequence of its natural motion, which is that of continuous revolution.

424. As they would have to do if motion along some particular semi-circle was indeed an intrinsically-determined motion for the revolving body.

425. cf. *in Cael.* 1.9, 292,1-7.

426. Simplicius here talks as though heaviness is not a constant feature of a given quantity of some particular stuff (hence it is not assimilable to Newtonian mass: but cf. n. 439 below), but rather varies according to the particular spatial location of the object in question; and this is indeed suggested by Aristotle's own lemma (see Hankinson and Matthen, forthcoming, ad loc.). But this is apparently different from the assumptions underlying the earlier arguments concerning the impossibility of there being an infinite body (*Cael.* 1.6), which turned (among other things) upon the premiss that an infinite body (irrespective of its position, or its state of motion or rest) would have to be infinitely heavy (or infinitely light), where weight and lightness are apparently conceptualized as constant potentialities of a given mass. But the case is not at all clear: at *Cael.* 4.3, 310a32-b1, Aristotle writes: 'if, then, what causes movement upwards and downwards is what causes weight and lightness, while what is moved is potentially heavy and light, the movement of each thing towards its own place is movement towards its own form.' This suggests that the weight of each object is a function of its size, its type, and its spatial location; and each object is only fully heavy (or light) to the extent of its capacity to be so when it has arrived in its proper place; but that is not the only possible explanation of the text.

427. Again the train of thought is not lucid here: Simplicius seems to argue that, because bodies cannot be increased (sc. in mass or volume?) to infinity, then neither can their associated weights or lightnesses (which is to suppose that weight and lightness are direct functions of mass or volume, or some equivalent); but that since a body moving an infinite distance would acquire an infinite velocity, and since that is associated with infinite mass (or equivalent), and since the latter is impossible, then no body can accelerate to an infinite speed. But it is hard to acquit that argument of a confusion between the two manners in which one might suppose a weight to become infinite (n. 426 above).

On the other hand, he is doing his best with Aristotle's text as it stands; and it seems as though the argument could have been developed without any reference to the notion of infinite weight, simply in terms of the impossibility of infinitely fast motion. It should be noted that nothing that Aristotle says here or elsewhere rules out the possibility of an object's accelerating asymptotically to approach (but never attain) some terminal velocity (cf. *in Cael.* 223,36-224,3: n. 151); and it has nothing to say against the contention that, even if the relation between time of travel and velocity is linear, the velocity will simply increase without limit (for any determinate time after the starting time it will be moving with some determinate, non-infinite velocity): and this Simplicius accepts at 264,9-12 below.

428. Cf. *Meteor.* 1.7, 344b19-25 (the fact that comets presage drought is an indication of their hot, dry nature); and *Meteor.* 2.3, 359a12-14: (the fact eggs will float in salt but not fresh water is an indication of brine's greater density). Aristotle sometimes uses the term '*tekmêrion*' in a more restricted technical sense: signs qualify as *tekmêria* just in case they are the middle terms in sound first-figure inferences. Thus, lactation is a *tekmêrion* of pregnancy because everyone who lactates has conceived. By contrast, other signs of pregnancy, such as sallowness, are defeasible, since while it is true that all pregnant women are sallow, not all sallow people are pregnant (*An. Pr.* 2.27, 70a2-b6). But in all these cases, as Simplicius indicates here, the *tekmêria* are effects of some underlying cause, and hence indicative of those causes; see n. 429 below.

429. See n. 428 above.

430. See n. 426 above.

431. This suggests that when objects are moving unnaturally and forcibly, they should move more slowly the further away they are from their natural places (and hence the closer they are to their opposite places); but since their opposing drive towards their natural place decreases in proportion to the increase in their distance from it, it would seem that, for a constant force being applied, the objects should move faster the further away they are from their natural places (since their tendency to resist the imposed motion diminishes). Either Simplicius has not seen this, or he does not mean 'in an opposite manner' to mean 'with an opposing alteration in velocity'.

432. Aristotle holds that the nearer a body approaches its natural place, the more it actualizes its form: *Cael.* 4.3, 310a33-4 (cf. *in Cael.* 1.2, 20,10-17). The 'proper wholeness' may refer to this more completely actualized form in the moving body, or more probably to the mass of its congener-element which is already in its proper place (or conceivably to both).

433. This work does not survive. Hipparchus (*c.* 180 – *c.* 120 BC) was one of the greatest astronomers of antiquity. He corrected many of Eudoxus' observations, and wrote a commentary (his only surviving work) on the *Phainomena* of Aratus, a scientific poem (which also survives). He discovered the precession of the equinoxes, which he approximated to within 10% of their true value, and offered an estimate of the value of the tropical year which was accurate to within six minutes. But most importantly he inaugurated the mathematical model of the heavens, which was to be adopted and adapted by Ptolemy and remained the orthodoxy until the Copernican revolution, by replacing the Eudoxan (and Aristotelian) system of concentric, offset spheres with a geocentric system in which the planets, sun and moon were located on small orbits (epicycles) which themselves revolved around empty points on larger orbits (deferents) which in turn revolved around the earth. Refinements of this basic model were able to produce mathematical approximations to (and hence predictions of) the apparent positions of the heavenly bodies, and to account for such recalcitrant phenomena as the retrogression of planetary motion, with unprecedented accuracy. On this passage, see Wolff, 1988.

434. Thus Hipparchus anticipates the impetus theory of Philoponus and mediaeval physics: forcible motion occurs when a quantity of motive power is impressed into a mobile object, a power which gradually exhausts itself: see Sorabji, 1988, chs 14-15.

435. Alexander is thinking of, e.g., fire kindled on the surface of the earth: no forcible motion has brought it to where it is, although it is where it is unnaturally, and hence will immediately begin to move towards its natural place.

436. i.e. from the centre: see nn. 426-7 above.

437. Here Alexander is referring to natural meteorological phenomena such as that of precipitation (*Meteor.* 1.9-12): water is evaporated by the sun as a form of air, and hence rises (1.9, 346b20-31); in the upper air (at certain times) it cools, becoming water again and hence predominantly heavy, and hence falls (346b32-5), which process varies with the movement of the sun along the ecliptic (more evaporation in the summer, more precipitation in the winter: 346b35-347a7).

438. The thought is presumably this: the impetus for the movement of a balance comes from the heavier of the two objects in the pans (since the lighter one is not absolutely light, and will not drag the other pan upward); but the pan moves faster the further down it goes, so it is another example of the increase in heaviness consequent upon closer approach to natural place.

439. On this (unattributed) account, weight will remain a constant of a particular bulk (i.e. it approximates to the modern concept of mass); heavier things will still in general fall faster than lighter things, but their acceleration is to be attributed not to their increasing weight in descent, but rather to the decrease in bulk of the impeding substance.

440. i.e. the decreasing quantity of intervening air.

441. Because there will always be an infinite amount of underlying air.

442. Since there can be no such diminution: n. 441 above.

443. i.e. on the (counterfactual) supposition that the object is infinitely removed from its proper place, the only thing which could account for its more rapid movement would be an increase in its weight; but of course given the strictures on the unapproachability of the infinitely distant, this will not work either (it is possible that *peperasmenou*, should be read in place of *apeirou*, at 266,26 above, yielding the sense 'if, provided the distance is finite, etc.').

444. A rare (and intriguing) indication of empiricism on Simplicius' part; such distinctions are now of course determinable with sensitive gravitometers.

445. i.e. one would have to assume that the larger object always just happened to be moved by something even greater: but there is no reason to suppose that any such thing would occur.

446. Strato, who abandoned much of orthodox Aristotelian theory, rejected the notion that lightness is an intrinsic property: frs 50-2 Wehrli.

447. See Lucretius 2.184-216.

448. Strictly speaking, Epicurus thought that bodies move only downwards, there being no centre to his infinite universe, where 'upwards' and 'downwards' are determined locally: *ad Hdt.* 60-1; see 1.7, 243,17, n. 311.

449. As I construe the thought here, Simplicius first supplies another argument which Aristotle might have deployed, to the effect that it requires more extrusive power to extrude a large mass, but large masses of such things move faster than smaller ones, then suggests that he did not do so either because he already had enough arguments and wanted to avoid prolixity, or because he anticipated an objection to the argument: the large extruded mass is surrounded by a larger extrusive force, and hence in extrusion there is a proportionality between extruder and thing extruded, which is not always the case in other types of forcible movement. However an anonymous vetter suggested reading the text as follows: 'But he changed the argument because of its facility, and at the same time he has perhaps tacitly forestalled the objection which states ...; but thus is not necessarily the case for something which moves <another thing>' (thus making the last clause a reply to, and not part of, the objection, and hence the reason why the objection is forestalled); this is certainly a possible construal of the Greek, but on balance I think that my version makes better sense.

450. This is only true if 'by force' is interpreted to mean 'against their natural inclination' – which of course begs the question. A partisan of inertial physics (or for that matter an Epicurean) can perfectly well suppose that a force must be applied to alter the existing velocity of an object, without thereby supposing that that object must possess some intrinsic natural velocity (or motile tendency): see above, nn. 398, 401.

451. See 247,30-5.

452. i.e. towards other masses of the same type; Simplicius does not of course mean that the mass itself exerts the pull, but rather that, other things being equal, the natural place is where the bulk of the element for which it is natural will be. But the phrase is more than merely pleonastic, since the detached part will come to rest, and naturally so, when it abuts upon its congener material even if it would still move further were that material not there: see *in Cael.* 1.2, 22,18-29; and Hankinson, forthcoming 2, §IV. Elsewhere (*Cael.* 1.2, 268b26-69a2), Aristotle uses the expression 'those things akin to <the elements>', and Simplicius, following Alexander, interprets this to mean different types or concentrations of earth, etc. (*in Cael.* 1.2, 16,18-26), as well as allowing that water is 'akin to' earth, and air to fire. But that cannot, it seems, be the sense here.

453. Reciprocal motion, *antiperistasis*, is one of the central concepts deployed by continuum-theorists against the atomists; in particular it is designed to show how motion in a plenum is possible (something the atomists denied), and that motion does not require actually-existing void space: Aristotle *Phys.* 4.7, 214a27-b3; cf. ps.-Aristotle *MXG* 976b22-9; Strato fr. 63 Wehrli.

454. Two crucial principles are mentioned here: (a) that of *horror vacui* and (b) that of the impenetrability of solids. The former was employed by continuum-theorists in their rebuttal of atomism; the latter was a universally-accepted conceptual truth (employed for example in the demonstration of the existence of invisible pores in the skin on the basis of sweating: sweat passes from the inside of the body, but nothing can pass through a solid; so there are invisible pores: Sextus *PH* 97, 140, 142). In this case (a) explains (in some sense) why the lighter object can be sucked down into the space forcibly vacated by the heavier one; (b) why it must do so by moving around (and not through) it.

455. The text is corrupt here: I have translated *pheromenên tautên tên doxan*, the reading of b and c, in place of *pheromenên tautên* as printed by Heiberg, of which I can make no sense. But my construal is somewhat strained, and the text may require more serious surgery.

456. *Timaeus* 62C.

457. See n. 402 above.

458. i.e. is determined geometrically and not by reference (except perhaps incidentally) to some body; this view, while making better sense of the concept of natural place, is hard to reconcile, as Simplicius points out, with Aristotle's official definition of place.

459. Because there is nothing outside the outermost sphere to contain it; thus, strictly speaking, it would seem that, on Aristotelian assumptions, the universe as a whole can have no place. Aristotle himself grapples with this issue at *Phys.* 4.5; but, as Simplicius goes on to imply, with limited success.

460. One of the rare occasions in which the syncretizing Simplicius allows that Aristotle's position on anything is seriously inadequate.

461. *Phys.* 8.6, 259a6-20; cf. in general 8.8-9.

462. i.e. the unmoved mover. Simplicius, after the fashion of ancient interpretation, treats the Aristotelian corpus as an organic whole, every part of which ought to be consistent with every other; but it is in fact very doubtful whether Aristotle subscribed to the theory of the unmoved mover at the time he composed *Cael.*

463. *Metaph.* 12.8, 1074a31-8.

464. Two things can be numerically distinct but share the same form and definition only if their form is realised in distinct matter; but this cannot be the case for something incorporeal. The problem with this view is precisely that Aristotle envisions a plurality of unmoved movers: *Metaph.* 12.8; see 270,18-27 below.

465. For Aristotle, the Prime Mover causes motion by inspiring in everything else the desire to emulate as far as possible its perfect actuality: *Metaph.* 12.7, 1072a26-b1

466. It is hard to see how this is supposed to answer Alexander's worry.

467. *Metaph.* 12.8, 1073a23-34.

468. Here Simplicius interprets Aristotle through his Neoplatonist lenses: the world of Plotinus and his followers is a hierarchical structure beginning with the single, unified, ineffable One, from which are generated in order Mind, Soul, and finally the visible cosmos: cf. e.g. *Enneads* 5.1, 5.2, 5.4.

469. Here Simplicius uses the term *aithêr*, in un-Aristotelian fashion, to mean 'fire'.

470. *Metaph.* 12.8.

471. Here again Simplicius expresses the characteristic Neoplatonist obsession with preserving the unity of the whole.

472. Here Simplicius takes sides in a debate on the nature of Aristotle's Prime Mover which persists to this day (cf. *in Cael.* 1.4, 154,6-156,24). While the treatment in *Phys.* 8.10, 266a12-b27 appears to make the Prime Mover an efficient cause, the case of *Metaph.* 12 is less clear. All that Aristotle explicitly affirms there is that the PM is a final cause (although some have seen reference to causal efficiency in 12.7, 1073a5-11), and interpretative orthodoxy has sided with Alexander against Simplicius here; however see most recently S. Broadie, 'Que fait le premier moteur d'Aristote?', *Revue Philosophique* 183 (2001), 375-411.

473. *Cael.* 1.5, 271a33. The slogan 'nature does nothing in vain' and derivatives is extremely common in the Aristotelian corpus (e.g. *PA* 2.13, 658a9; *GA* 2.6, 744a36); but this is the only occasion on which any mention of God is appended, a fact which suggests that *Cael.* may be an early work. At all events, it seems foreign to the purely intellectual conception of God Aristotle develops in *Metaph.* 12 and *NE* 10.6-9 that his God should actually do anything.

474. Ammonius' treatise proving that Aristotle's God is indeed an efficient cause does not survive; but Simplicius summarizes it at *in Phys.* 1361,11-1363,12; see Introduction.

475. Here again Simplicius betrays his Neoplatonist leanings in favour of the absolute, transcendent, original One at the top of the hierarchy of being; cf. the extended discussion of Neoplatonist metaphysics at *in Cael* 1.3, 92,32-107,24.

476. Reading *pros tois oikeiois topois*, with A and D, against Heiberg's *pros tous oikeous topous*, which would mean 'towards their proper places', and gives the wrong sense; see 263,13-31 above.

477. The word used here, *anelittein*, might simply mean 'revolving' (LSJ, s.v.); but in Aristotelian cosmology, it generally has the specific sense of 'moving backwards', i.e. the counteracting spheres introduced by Aristotle to give a physical realisation to the purely mathematical models of Eudoxus and Callippus (cf. *Metaph.* 12.8, 1074a1-12; and see *in Cael.* 1.2, 32,16-22). The point is that, for Aristotle, these extra spheres also need to be counted, since the figures for either the Eudoxan or the (slightly more complex) Callippean system could not give the total number of required unmoved movers. On the systems of Eudoxus, Callippus and Aristotle, see Dicks, 1970, chs 6-7.

478. Here Simplicius refers to the problem created for the Aristotelian notion that there are four sublunary elements by the fact that there are, directionally speaking, only two rectilinear natural motions: yet natural motions are supposed to differentiate the elements. Aristotle will himself on occasion talk of three elements, rather than five: (*Cael.* 3.1, 298b6-8), as Simplicius notes (*in Cael.* Prologue, 1,18-24; 3.1, 555,7-12; and 1.6, 216,13-15: n. 96).

479. i.e. earth.

480. Here Simplicius supposes that earth is *sui generis*, while fire, air and water are three parts of the remaining genus; this is odd, since elsewhere water is explicitly held to be heavy (although less heavy than earth), and hence to form a natural class with earth, leaving the pair of air and fire as the light elements (*in Cael.* Prologue, 1,18-24).

481. Alexander had suggested that 'outside' (*Cael.* 1.8, 277b18) did not mean 'outside the cosmos', but simply 'outside its proper place' (272,20-33); Simplicius rejects this interpretation on the grounds that Aristotle immediately goes on to say that it is impossible for the elements to be 'outside', a claim which will be true for him only if 'outside' means 'outside the cosmos'.

482. i.e. relative to its proper position in this world: see 252,9-254,5 above (ad *Cael.* 1.8, 276b14-18).

483. Retaining *ho* of the MSS, deleted by Heiberg.

484. 272,21-33.

485. cf. *Cael.* 1.2, 269a33-4; *in Cael.* 1.2, 50,7-17; 1.7, 276a12-15; *in Cael.* 1.7, 243,30-244,33.

486. i.e. intermediate between the central place of the heavy body (earth) and the outermost place of the heavenly ether: Simplicius still operates under the assumption that the 'intermediate' three bodies (fire, air, water) are generically similar in that they all possess lightness to some degree: see above, nn. 478, 480; but see also n. 487 below.

487. In this parenthesis, he treats 'the rising body' as naming fire alone (although of course air has a tendency to rise); air and water are, of course, in another sense intermediates, intermediate between the absolutely heavy and the absolutely light (*Cael.* 4.4-5). But it seems clear in the lemma (*Cael.* 1.8, 277b13-23) that Aristotle intends by 'intermediate' here the region of any bodies which have a tendency to rise (i.e. fire and air), with the 'place around the middle' referring to the place of both heavy elements, water as well as fire; here as elsewhere the indeterminacy in the meaning of 'intermediate' is an irritation.

488. i.e. *Cael.* 3-4.

489. The perceived difficulty with calling the ether an element is, presumably, that it is central to Aristotle's element-theory that elements intertransmute (*Gen. Corr.* 2.3-4), while the ether of course does not admit of any alteration; moreover, Aristotle also says quite regularly (e.g. *Gen. Corr.* 2.3, 330a30) that there are four elements, thus implicitly excluding the ether. But the scope of *Gen. Corr.*, as well as of *Cael.* 3-4, is, as its title suggests, explicitly that of the sublunary elements; and certainly the methodology of *Cael.* 1.2, associating the fundamental elements with the three primary motions (of which one, the circular, is the natural motion of the ether), seems clearly to imply that, for Aristotle, the ether is, in a perfectly good sense, an element.

490. *Cael.* 3.1, 298b6.

491. *epistêmonikôs*, in the manner of an *epistêmê*, a rigorously-organized body of knowledge, or science. For Aristotle, the objects of an *epistêmê* cannot be other than they are: *epistêmai* deal with eternal, necessary truths (cf. *NE* 6.3, 1139b18-24), and contingent facts are not (as such) the object of science (*Metaph.* 6.2).

492. *Cael.* 1.10-12.

493. Simplicius is referring, obliquely, to Aristotle's discussions in *Metaph.* 7.6 and 7.10-11 here, in particular 7.10, 1035b32-1036a12; also relevant is *Metaph.* 7.15 (esp. 1040a28-b4), where he argues that even in cases where the species is constituted by a single individual (as in the case of the sun and the moon), there is still no definition of the individual as such; rather the definition is of the (uniquely instantiated) form.

494. Again, the discussion of *Metaph.* 7.15 is relevant here.

495. The fifth-century Pythagorean Philolaus of Croton elaborated a somewhat obscure metaphysical theory in which the world was composed of 'limiters' and 'things limited' (44 B 1-3, 6 DK), in which the limiters are presumably formal, and hence general and separate; but Simplicius may well be thinking of a much later tradition of Pythagoreanism, associated with such figures as Numenius and Moderatus of Gades, a tradition in which 'genuine' Pythagorean thought is hard to disentangle from later, Platonizing accretions (for comparisons between Platonism and Pythagoreanism in this regard, see *Metaph.* 1.5-6; 1.8, 989b29-990a32). The attribution to Socrates of a belief in transcendent forms betrays Simplicius' syncretistic commitments (Aristotle rightly sees him as being concerned only with definition, and then only in the case of ethics: *Metaph.* 1.6, 987b1-4). For Plato, the classic texts on the theory of the independently-existing, transcendent Forms include *Phd.* 65D-67B; 74A-75D; 78C-79D; 100B-105B; *Parm.* 128E-135D; *Rep.* 5.474C-480A; 6.506B-511E; 7.514A-541B; *Tim.* 27C-29D; *Symp.* 210A-212B. *Phil.* 16C-27C propounds a theory involving limits and the unlimited that in some respects parallels that of Philolaus.

496. The latter view, that form is something only exemplified in matter, and hence that there are no such things as separated, self-subsistent Platonic Forms, is of course Aristotle's: cf. his criticisms of Plato at *Metaph.* 1.6; 1.9; 3.6; 7.2; 13.4-5; *NE* 1.6. Note that Simplicius' syncretism does not blind him to this important distinction.

497. The word translated 'species' here, *eidos*, to contrast it with genus, is the same word Aristotle standardly uses for form.

498. The issue here turns on the acceptability of the notion of particular, individual forms, a vexed question in Aristotelian scholarship. In one sense, the form of Socrates is clearly that of 'human', and as such is multiply instantiable. But some have argued that Aristotle both requires, and is committed to, a stronger sense in which the form is the form of the individual (see Frede, 1985). On that supposition, which Simplicius appears to countenance here, the question of the generality of that form becomes acute: is it analogous to a genus or a species? If so, it would seem that it ought to be multiply instantiable; but it is *Socrates'* form – and hence that seems absurd. Hence, we are invited to conclude, either Socrates' form is not a species or a genus (in which case it will not follow that all forms are potentially multiply instantiable), or alternatively that something can be specific (or even generic) without that fact implying multiple instantiability: and on either alternative Alexander's objection fails.

499. i.e. the Forms of the Platonists.

500. i.e. the distinction between the two alternative ways of considering the nature of forms, whether as separately existing and self-subsistent or not (see nn. 495-6 above): in both cases it seems that many individuals will share a form.

501. *Tim.* 31A-B: see 286,20-287,14 below.

502. cf. *NE* 1.6, 1096a34-b2, where Aristotle argues that 'Man-in-itself', the Platonic locution, does no work that is not already done by the universal 'man'.

503. i.e. the Forms of the animals in question.

504. cf. *Metaph.* 7.15, 1040a28-b4; that the heavenly bodies are alive is emphasized at *Cael.* 2.12, 292a18-b2, although their motion is not caused by soul (as Aristotle had supposed in his early, Platonizing days: Cicero, *On the Nature of the Gods* 2.15.42, 2.16.44 = frs 23-4 R).

505. The 'image' here is the general idea of the universal which exists in the Soul of the world, and which answers to the purely intellectual forms of the Mind; here Simplicius intrudes his own rather baroque Neoplatonism into the discussion. For the idea that Mind, which itself emanated from the One, then produces in turn Soul which itself generates these 'images', see Plotinus 5.2.1,14-19.

506. i.e. Platonic transcendent Forms; these images, then, answer (as they standardly do in Neoplatonist syncretism) to Aristotelian immanent forms or universals, single sets of characteristics held in common by each member of a species.

507. i.e. concrete physical particulars.

508. Here Simplicius is apparently gesturing towards astrology, and in particular to the uniqueness of each individual's horoscope, which accounts (he supposes) for the very different fates of different individuals of the same general kind. On the prevalence of astrological beliefs in later antiquity, in particular among Stoics and Platonists, and on the sceptical arguments directed against it, see Dodds, 1951, 245ff.; Long, 1982; Hankinson, 1988; and Hankinson, 1998, VIII.5.a-b; XI.1.b; XII.2.d.

509. *Tim.* 31A-B; see further below, 286,20-287,14.

510. This sentence is Simplicius' reply to Alexander's argument; it supplements Plato's argument of *Tim.* 31A-B.

511. i.e. if the lower exemplars yield many instances of themselves, the highest one ought to do so to as well, and to a greater extent.

512. This sentence is again Simplicius' response; in effect he again adopts the principle, familiar from a certain sort of *a priori* theology, that the logic of perfection demands uniqueness; see below, n. 592.

513. Simplicius' reply once again: here the emphasis is on the Neoplatonic principle of intelligibility that demands that, if there is a multiplicity, it must be accounted for in terms of some prior underlying unity; cf. Proclus, *Elements of Theology* 1.6, 11-24.

514. This sentence is omitted from the summary of Simplicius' lemmas in Heiberg: Simplicius glances at it at the end of this discussion. It might make better sense to end this passage of discussion at 279,25, placing 279,25-30 at the beginning of the next passage, and attaching the last sentence of this lemma

(278b9-11) to the beginning of the next; but there is insufficient justification for changing the order of the text as it stands.

515. i.e. the first premiss, that the heavenly form is distinct from its composite.

516. 'Or the snub' may well be an intrusion here: Aristotle frequently uses snubness as an example of form in matter (e.g. *Metaph.* 6.1, 1025b30-1026a6; 7.5, 1030b14-35); but he does not do so in this passage, where, uniquely, and for no very obvious reason, he prefers the aquiline.

517. i.e. serially. Simplicius implicitly rejects a purely material criterion of object-identity here (i.e. two distinct objects can share all their material parts, and the arrangements of those parts, in common, and yet remain genuinely distinct), although it is not clear on what grounds he would do so (compare the Epicurean arguments to dispel what they take to be metaphysically unfounded fears of life after physical dissolution: Lucretius, 3.830-911).

518. See n. 498 above.

519. i.e. those things whose very definition involves a certain type of matter, as for example snubness is concavity *in a nose*: cf. *Metaph.* 7.10, 1035a4-6, 22-30.

520. *Cat.* 5.

521. Or 'in a subject': in the *Categories*, *hupokeimenon* has a more grammatical sense than it does elsewhere in Aristotle's more overtly metaphysical writings.

522. i.e. the composite is not in a substrate, since, being *enmattered* form, it already involves the substrate.

523. Heiberg refers to *de An.* 2.1, 412a22-412b1 (cf. 412a9-11), on the distinction between first and second actuality, exemplified by the example having knowledge but not using it and both having and using it. But that distinction is not precisely to the point here, which rather concerns separable and inseparable complete actualities, as the next line makes clear: see nn. 524-5 below.

524. *De An.* 2.2, 413b25-7.

525. *De An.* 3.5, 430a17-18; in this notoriously difficult chapter, Aristotle sketches his account of the active intellect, the only part of the mind which is (in some, highly controversial, sense) separable from physical instantiation.

526. The distinction made here is not entirely clear: presumably Simplicius sees Aristotle as distinguishing between thinking of the outermost orbit in purely formal terms, and considering it as a concrete physical realisation (cf. 277b27-278a16 above); this being the case, 'essence' might be a better rendering of *ousia* in this sentence than 'substance'.

527. *Cael.* 1.3, 270b6-7.

528. Alexander's point is that Aristotle must not be supposed to confuse what he habitually calls the upper place, i.e. the natural place of fire immediately beneath the lunar sphere, with what he here calls 'uppermost', i.e. the highest reaches of the universe as a whole. Aristotle nowhere precisely defines the upper place (the natural place of fire) as being the innermost surface of the lunar sphere, and his use of 'extremity' is rather loose; but at *Cael.* 4.3, 310b8-9, he writes that 'since place is the limit of the containing body, and the extremity and the middle contain everything which moves up and down', which strongly suggests the definition in question; see further 4.1, 308a13-33; 3.2, 300b17-26.

529. i.e. the spheres of the sun, the moon, and the planets (the 'some of the stars' of 278b17-18).

530. i.e. the place of the heavenly bodies (taken as a whole), not that of fire: see n. 528 above.

531. i.e. it is in virtue of the supreme divinity of the outermost sphere (and presumably, in Simplicius' view, its unmoved mover) that the lower spheres inherit their share of divinity. The verb *metekhetai*, here translated as 'are shared in', is the passive form of *metekhein*, to participate or share in, the technical Platonic term for the relation that holds between particulars and Forms; this, therefore, names the converse relation, that between Forms and particulars, as is standard in Neoplatonism.

532. The planets again: see n. 529.

533. *Cael.* 1.2-4.

534. *Cael.* 1.10-12.

535. *Pol.* 269D.

536. *Tim.* 28B.

537. Plato fancifully derives the word 'man' (*anthrôpos*) from *anathrôn ha opôpen*, 'looking up at what he sees': *Crat.* 399B-C.

538. i.e. man alone is capable of forming universal concepts (this view was not universal in antiquity). It is not clear what etymology Simplicius refers to: perhaps something like *eis hen athroizôn*, 'collecting into one'; but Simplicius does not employ the verb *athroizein* here.

539. Plato (*Crat.* 396C) actually derives the name of Ouranos the god from *tou horan ta anô*, 'looking upwards'; but he evidently had the heavenly bodies in mind (the planet Uranus was of course unknown to the ancients).

540. i.e. since the whole of the universe depends both efficiently and ontologically on the supremacy of the heavenly movements (cf. *Cael.* 2.12; *Gen. Corr.* 2.10; *Phys.* 8.8-10), it is reasonable to designate the whole thing thus.

541. Or possibly, reading *to sôma tês hulês* with E^2, 'of all the matter there is'; or perhaps read *tês sômatikês hulês*, 'of all the material body'.

542. See *Cael.* 1.2, 268b14-269b13.

543. *Cael.* 1.4.

544. *Cael.* 1.8, 276b18-21; cf. 276a30-b18; and see *in Cael.* 1.8, 250,17-254,25.

545. i.e. in this world.

546. See *Cael.* 1.8, 276a18-b21.

547. In this case, exhaustive disjunctive syllogism: see *in Cael.* 1.7, 227,28-228,13: nn. 190-1, 193.

548. The text is clearly deficient here: this translates Heiberg's plausible supplement.

549. See n. 544 above.

550. *Cael.* 1.7, 276a12-15; *in Cael.* 1.7, 243,30-244,24; cf. *Cael.* 1.2, 269a32-269b2 (on motion).

551. As apparently Aristotle does at 279a6-9.

552. See the previous lemma: 278b25-279a2 (and cf. *Cael.* 1.8, 276a30-b21; *in Cael* 1.8, 250,17-254,25).

553. In Aristotle's developed theory of scientific explanation in *Post. An.*, a demonstration is a syllogistic deduction from necessary premisses, in which the middle term is the cause of predicate's holding of the subject in the conclusion, and hence serves to explain it. Thus, as the next sentence shows, although in cases where the terms of the major premiss convert, two sound syllogistic inferences are constructible, in which the conclusion of one is the minor premiss of the other, only one of them will be explanatory. In Aristotle's example, given that (of heavenly bodies) all and only those which are near do not twinkle, we may infer that planets are near because they do not twinkle or that they do not twinkle because they are near: but only the second inference embodies a genuine explanation, since it is their nearness which causes their not twinkling, and not *vice versa*. *Post. An.* 1.13, 78a23-b4 (and cf. 2.17).

554. This inference (and the subsequent ones) can be represented in the second figure form Camestres (see 1.7, n. 252): A = it is possible for there to be a body; B = place; C = outside the heaven.

555. Again this argument is in Camestres: A = being empty of body but capable of receiving it; B = void; C = outside the heaven.

556. *kinêsis*, translated elsewhere as 'motion', but Aristotle's definition of time clearly demands the more comprehensive rendering; for the definition, see *Phys.* 4.11 219b3; 4.12, 221b7-23.

557. This complex argument is multiply unsatisfactory. The first inference is valid enough (if there is motion, there is time; if there is a moving body, there is a motion; hence if there is a moving body, there is time); but the second inference invokes the notion of body *tout court*, not that of a moving body, and hence cannot makes use of the conclusion of the first, along with the further premiss that no body exists outside the heaven, to drive the conclusion that there is no time there. Moreover, even if we allow that there is no moving body outside the heaven (relying on the impossibility of one *coming to be* there), the inference is still invalid. Effectively, Simplicius requires not the first inference as it is stated, but rather its converse (reversing the implications of both premisses) which yields the conclusion that if there is time there is a moving body; which, conjoined with the premiss that there can be no *moving* body outside the heaven, in turn yields the conclusion that there can be no time there. Matters could be cleared up by altering the received text – but the emendations would have to be severe, and it seems likely that Simplicius nodded here.

558. The Stoics' doctrine of extra-mundane void is well-attested: Diogenes Laertius 7.140; cf. *SVF* 2.502-3. This passage is printed as *SVF* 2.535. The argument which follows is also attributed to Archytas (see Sorabji, 1988, 125-8, 135-6), and was employed by the Epicureans to demonstrate the inconceivability of a finite, bounded universe: Lucretius 1.958-97. The Stoics had a physical reason for their postulation of the void, namely that when the world was resolved utterly into fire (as, on their view, happens from time to time) it must, fire being rare, take up more room than it now does. So there must now be more room for it then to take up (Cleomedes 6,11-17 = *SVF* 2.537); see n. 566 below.

559. And so in either case there will be something beyond the supposed boundary.

560. Alexander's argument also begs the question: for if there is something (even void) outside this world, then this world is not identical with the universe as a whole, something the Stoics explicitly acknowledged: 'they say that the world is the whole, but the external void together with the world form the universe' (Sextus Empiricus *M* 9.332 = *SVF* 2.524).

561. *Tim.* 63B.

562. *Tim.* 63C.

563. *Cael.* 4.3, 310b3-4: Aristotle actually supposes earth to be moved to the location of the moon.

564. But see n. 560 above. Alexander also dealt with this issue at *Quaestiones* 3.12 (see Sorabji, 1988, 136-8); and cf. Alexander in Simplicius *in Cael.* 467,1-3.

565. This was a general ancient thesis, upheld by Aristotle (*Cat.* 7, esp. 6b29-8a12) and apparently endorsed by the Stoics (Simplicius *in Cat.* 7, 1666,15-29 = *SVF* 2.403), although it requires care – a son may outlive his father (and indeed *vice versa*); Alexander shows his awareness of this in the following lines: 'either is *or could be* something which is capable of being received.'

566. Two issues seem to be conflated here. Alexander's main point seems to be that, if the extramundane void is infinite (as Chrysippus held: Stobaeus

1.161,19-26 = *SVF* 2.503), then if it is to be 'capable of receiving' body (that the definition is Stoic is confirmed by Cleomedes 8,10-14 = *SVF* 2.541; Sextus *M* 10.3 = *SVF* 2.505) as a whole, there must be an infinite body for it to receive – but there is, as Alexander says, on the Stoics' account, no such body: for them, the totality of material stuff is finite (cf. Sextus *M* 9.332 = *SVF* 2.524). But of course this objection does not apply to the totality of finite parts of the extra-mundane void, taken distributively – any part of the void is such as to be (potentially) receptive of body. But the second claim may well be that none of this void will ever, as a matter of fact, be occupied by any body (since this world contains all the matter there is, and no matter from this world, for the reasons extensively canvassed in this and the previous chapter, can ever leave its confines). Consequently, on the basis of Aristotelian physical considerations (which the Stoics reject: they hold that the volume of the world, while always finite, is still variable, which is why they need the extramundane void in the first place: n. 558 above), supplemented by the modal doctrine that what is possible must at some time be actual, the possibility of the extramundane void can be discounted. But of course the Stoics need not (and did not) accept those presuppositions: see n. 568 below.

567. An unorthodox Peripatetic philosopher of the first century BC, tutor of the geographer Strabo, and associate of Arius Didymus and the emperor Augustus (Strabo 14.5.4). He wrote a text *Against Aristotle on the Fifth Substance*, substantial and interesting portions of which are preserved in Simplicius' commentary on *Cael.* 1.2 (see Hankinson, 2002; and see also Hankinson, forthcoming 2).

568. In other words, if this infinite external space is 'receptive' of body, it should at some time receive it (or at least be *capable* of receiving it: i.e. nothing conceptually stands in the way of the supposition), which is contrary to the Stoics' hypothesis. But Xenarchus' suggestion is not as otiose as Alexander suggests, and it looks back to a substantial and influential dispute on the nature of modality between Diodorus Cronus and Philo of Megara in the late fourth century BC. Diodorus defined the possible as 'what either is or will be the case' (Epictetus *Discourses* 2.19.2-5), Philo more generously as 'what is predicated in accordance with the bare suitability of the subject' (Alexander, *in An. Pr.* 183,34-184,10). A Philonian can hold something to be possible even if it never will occur (and perhaps if it never could occur): thus wood on the sea-bed is still inflammable (Alexander, loc. cit.). Xenarchus' linguistic revision is designed to suggest that, while from a Diodoran perspective, the extramundane void is 'incapable of receiving' any body, none the less it is still, in the Philonian sense, receptive of it. Chrysippus apparently combined the accounts to yield the definition of the possible as 'that which admits of being true and which is not prevented by external circumstances from being true' (Diogenes Laertius 7.75; see Bobzien, 1998, 3.1); it is a further (complex) question whether on this definition it will be possible for (some remote part of) the extramundane void to be occupied (on Stoic physical suppositions) even if as a matter of fact it might never be.

569. i.e. it is not just that it is no part of God's function to ensure the preservation of the world: rather, since the Stoics identify God with fire, God must possess his own elemental nisus. The Stoics maintained that world was held together by the internal dynamic tension (see n. 571 below) of its all-pervading *pneuma*, which was, for them, to be identified with the divine: Diogenes Laertius 7.138-9.

570. i.e. having fire's natural upward tendency.

571. The Stoics held that the coherence of physical bodies was the result of their being permeated by *pneuma*, a sort of fiery air in a state of constant dynamic tension, which was also, in their system, the bearer of intelligence: *SVF* 2.439, 2.444, 2.449, etc.

572. cf. Alexander, *On Mixture* 223,25-224,26: some bodes are naturally cohesive, others naturally fluid – but it is otiose to explain cohesion in terms of the action of some other physical stuff (Galen makes a similar point: *On Containing Causes* 6.3).

573. i.e. the thought-experiement of the extension of the hand beyond the edge of the world: 284,28-286,2.

574. *Tim.* 31B; see above, 276,10-13, 277,17-19.

575. 276,29-277,19 above.

576. *Tim.* 32C-33A.

577. *Tim.* 33A-B.

578. The MSS of Aristotle, followed by the editors, here give *amaurôs*, faintly or obscurely; but Simplicius seems to have read *amudroteron*, more obscurely (which may be a misprint for *amauroteron*: cf. Simplicius, *in Phys.* 1361,22; Philoponus, *in Phys.* 189,15); and Themistius in his paraphrase offers a comparative form (*obscurius*: *in Cael.* 56,33); see further below, 288,23, 291,1, and n. 608. The sense is not affected.

579. Simplicius apparently interprets Aristotle's Greek in this manner (291,4-5); but it is more plausibly taken to mean 'the divine is wholly unchangeable': see n. 610 below.

580. *Phys.* 4.5, 212a31-b22; given Aristotle's own definition of place (see next clause and n. 581), it is impossible for the universe as a whole to be in place.

581. *Phys.* 4.2, 209a31-b1; 4.4, 211b6-14, 212a6-7; cf. *in Cael.* 1.8, 258,2-3 above.

582. *Tim.* 33A; above, 287,11-14.

583. Because they are encompassed by a further body, that of the fixed stars.

584. *Cael.* 1.9, 279a24: i.e. its maximal possible natural life-span. The concept of maximal natural potentialities will become important in the next chapter.

585. *Iliad* 24.725.

586. i.e. at 279a25-8 Aristotle etymologizes *aiôn* as deriving from *to aei einai* (presumably in the form *to aei on*).

587. See n. 578 above.

588. Perhaps the reference is, as Heiberg says, to *Metaph.* 12.7, 1072b13-30, esp. 28-30, where Aristotle says 'so we say that God is a living thing, eternal and most good, and consequently life, and an age [*aiôn*] which is continuous and eternal belong to God' – but that hardly justifies Simplicius' paraphrase in the next phrase. But compare *Metaph.* 12.9, 1075a6-10: 'everything immaterial is indivisible; and just as the human mind ... is in a certain period of time, so too is this very thought of itself throughout all the ages (*ton hapanta aiôna*).' 'Prior to this' is intriguing: Simplicius treats all of Aristotle's works as a unified whole, and thinks of them as having a natural ordering (part of which is discussed in the Prologue to *in Cael.*: 2,18-3,12); and in such an ordering the very general concerns of *Metaphysics* precede the more specialized treatise such as *Physics*, *De Caelo*, and *Meteorology*; and this may be all he intends with this reference to priority. But he may also have believed that *Metaphysics* was in fact composed earlier as well.

589. cf. *Tim.* 37C-D.

590. The latter are the works generally known (following their editing publication in the first century BC by Andronicus of Rhodes) as 'esoteric', i.e.

intended only for the audience of members of Aristotle's own school (on the history and organisation of which see Lynch, 1972).

591. Fr. 16 R.

592. The sort of *a priori* considerations attributed to Aristotle here are staples of theological discourse (see n. 512 above); they date back at least to Xenophanes (frs 21 B 23-6 DK).

593. *Rep.* 2.380D-E.

594. *Rep.* 2.381B-C.

595. *Rep.* 2.381C. Plato is arguing against the traditional account of the gods of Homer, which has them metamorphosing into various shapes, as well as engaging in lying, trickery, deceit, theft, and adultery, all of which Plato considers inconsistent with the notion of the divine. Such arguments again originate with Xenophanes (frs 21 B 11, 14-16 DK).

596. Simplicius (or perhaps Alexander: n. 597 below) is apparently quoting *Cael.* 1.9, 279b1-2 here; but in our MSS of Aristotle there is no 'naturally' – either Simplicius added it himself, or it appeared in his MS, or he mistakenly supposed that it did. Nothing of importance turns on this.

597. All of the previous discussion is apparently derived, although not directly, from Alexander.

598. i.e. the unmoved movers: cf. 1.8, 270,2: n. 462; see *Metaph.* 12.8 for the multiplicity of particular unmoved movers, one for each heavenly sphere; *Metaph.* 12.7 for the single, Prime Unmoved Mover.

599. *Cael.* 1.9, 279a11-12.

600. 279a17.

601. 279a18.

602. Seclude *ta* before *ourania.*

603. Simplicius, since he wishes to discern the Prime Mover in *Cael.*, sees in this phrase (*kuklôi pheromenon*) an implication of passivity, of being moved rather than simply being in motion (as opposed to the more frequent *kuklôi kinoumenon*, which I translate neutrally as 'moving in a circle', and the term used in the next line, *peripheresthai*, 'revolving'), which would of course suggest that something else is causing the movement; but it is not clear that it need always have this connotation.

604. *Cael.* 1.8, 277b16.

605. The MSS of Simplicius all read *dê*, 'indeed', instead of *de*, 'and', here (279a25): but the MSS of Aristotle all preserve '*de*', as do those of Simplicius when the quotation is repeated a few lines down (290,31); but this is probably a slip of the pen on Simplicius' part here, and hence I have not emended the text.

606. See above, n. 605.

607. 279a25-7.

608. Reading *exêrtêtai*, against Heiberg's *epeisin*, which has better MS authority, but which does not correspond with Aristotle's text; however this may be a deliberate correction or explication on Simplicius' part.

609. See n. 578 above. 'Precisely' and 'obscurely' here amount roughly to 'directly' and 'derivatively'.

610. See n. 579 above.

611. None of these claims is secure: Aristotle may simply have meant by 'first and highest' the outermost sphere. It is in any case highly controversial whether, when he wrote *Cael.*, he entertained any notion of the unmoved mover: and this passage, which is sometimes cited, following Simplicius, in favour of the view that he did, is inconclusive: see n. 603 above.

612. Simplicius again assumes that the doctrine of the unmoved mover is already in play for Aristotle here.

613. Heiberg fails to mark this clause as a quotation; and see nn. 579, 610 above.

614. See below, 291,24-8.

615. The grammar here is somewhat strained, but this appears to be the sense. The idea is that the 'it' of the 'So it is reasonable' sentence appears to refer back to the subject of the previous sentence, which appears to be the unchangeable divine; but this yields a hopeless contradiction, spelled out further in the next sentence, which may be summarized in the following apparently inconsistent triad: motion is a form of change; the highest being does not change; the highest being moves unceasingly.

616. i.e. to the subject-matter of the preceding sentences, 279a30-5.

617. i.e. Simplicius knew of MSS reading the active *kinei* ('causes motion'), in place of the middle or passive *kineitai* ('is moving' or 'is moved').

618. i.e. is absolutely unchanging.

619. i.e. for being moved: the Prime Mover first of all and directly induces movement in the *primum mobile*, or first moved thing, and only indirectly in all subsequent movers: cf. *Cael.* 2.12.

620. *Phys.* 8.8, 261b27-265a12.

621. Language recalled by T.S. Eliot in 'Four Quartets':
'In my beginning is my end' ('East Coker')
'What we call the beginning is often the end
And to make an end is to make a beginning.
The end is where we start from.' ('Little Gidding')

Bibliography

Arnim, H. von (1903-5) *Stoicorum Veterum Fragmenta* [*SVF*], 3 vols (Leipzig), (1924), vol. 4, indexes, by M. Adler (Leipzig)

Blumenthal, H.J. (1978) '529 and its sequel: what happened to the Academy?' *Byzantium* 48, 369-85 [and reprinted in his Blumenthal, 1993]

——— (1979) 'Themistius, the last Peripatetic commentator on Aristotle?' in Bowersock, G.W., Burkert, W. and Putnam, M.C.J. (eds) *Arktouros*, Hellenic Studies presented to Bernard M.W. Knox (Berlin and New York), 168-82

——— (1993) *Soul and Intellect: Studies in Plotinus and Later Neoplatonism* (Aldershot)

Bobzien, S. (1998) *Determinism and Freedom in Stoic Philosophy* (Oxford)

Broadie, S. (2001) 'Que fait le premier moteur d'Aristote?', *Revue Philosophique* 183, 375-411

Cameron, A. (1969) 'The last days of the Academy at Athens', *Proceedings of the Cambridge Philological Society* 195, n.s. 15, 7-29

Charlton, W. (1970) *Aristotle: Physics I and II*, Clarendon Aristotle Series (Oxford)

Dicks, D.R. (1970) *Early Greek Astronomy to Aristotle* (Ithaca NY)

Diels, H. and Kranz, W. (1951^6) *Die Fragmente der Vorsokratiker*, 3 vols (Berlin)

Dillon, J.M. (1973) *Iamblichi Chalcidensis in Platonis Dialogos Commentariorum Fragmenta* (Leiden)

Dodds, E.R. (1951) *The Greeks and the Irrational* (California)

Foulkes, P. (1992) 'Where was Simplicius?', *Journal of Hellenic Studies* 112, 143.

Frantz, A. (1975) 'Pagan philosophers in Christian Athens', *Proceedings of the American Philological Society* 119, 29-38

Frede, M. (1985) 'Substance in Aristotle's *Metaphysics*', in A. Gotthelf (ed.) *Aristotle on Nature and Living Things* (Bristol and Pittsburgh), repr. in Frede, 1987

——— (1987) *Essays in Ancient Philosophy* (Oxford and Minnesota)

Glucker, J. (1978) *Antiochus and the Late Academy* (*Hypomnemata* 56) (Göttingen)

Guthrie, W.K.C. (1936) *Aristotle: On the Heavens*, Loeb Classical Library (London and Cambridge MA)

Haas, F. de (1997) *John Philoponus' New Definition of Prime Matter* (Leiden)

Haase, W. and Temporini, H. (eds) *Aufstieg und Niedergang der römischen Welt* [*ANRW*] (Berlin and New York)

Hadot, I. (1969) 'Die Widerlegung des Manichäismus im Epiktetkommentar des Simplikios', *Archiv für Geschichte der Philosophie* 51, 31-57

——— (1996) *Simplicius: Commentaire sur le Manuel d'Epictète*, Leiden

Hadot, I. (ed.) (1987) *Simplicius – sa vie, son oeuvre, sa survie*, Peripatoi vol. 15 (Berlin)

Hadot, P. (1990) 'The life and works of Simplicius in Greek and Arabic sources', in Sorabji, 1990

Hankinson, R.J. (1988) 'Stoicism, science and divination', in R.J. Hankinson (ed.) *Method, Medicine and Metaphysics* (*Apeiron* supp. vol. 21, 1988)

——— (1998) *Cause and Explanation in the Ancient Greek World* (Oxford)

——— (2002) *Simplicius: On Aristotle On the Heavens 1.1-4* (London and Ithaca NY)

——— (forthcoming 1) 'Mathematics and physics in Aristotle's theory of the ether', in C. Wildberg and A. Bowen (eds) *Essays on Aristotle's De Caelo*

——— (forthcoming 2) 'Xenarchus, Alexander and Simplicius on simple motions, bodies and magnitudes', *BICS* 43

——— (forthcoming 3) 'Natural, unnatural, and supernatural motions: contrariety and the argument for the elements in *de Caelo* 1.2-4'

Hankinson, R.J. and Matthen, M. (forthcoming) *Aristotle: On the Heavens I*, Clarendon Aristotle Series (Oxford)

Heiberg, I.L. (1894) *Simplicii in Aristotelis de Caelo Commentaria* (Berlin), *Commentaria in Aristotelem Graeca* [*CAG*], vol. VII

Hoffmann, P. (1987) 'Simplicius' Polemics', in Sorabji, 1987

Lameer, J. (1997) 'From Alexandria to Baghdad: reflections on the genesis of a problematical tradition', in G. Endress and R. Kruk (eds) *The Ancient Traditions in Christian and Islamic Hellenism* (Leiden)

Larsen, D.G. (1972) *Jamblique de Chalcis, Exégète et Philosophe* (Aarhus)

Long, A.A. (1982) 'Astrology: arguments pro and contra', in J. Barnes, J. Brunschwig, M.F. Burnyeat, M. Schofield (eds) *Science and Speculation* (Cambridge)

Matthen, M. and Hankinson, R.J. (1993) 'Aristotle on the form of the Universe', *Synthese* 96-3, 417-35

Moraux, P. (1965) *Aristote: Du Ciel* (Paris)

——— (1984) *Der Aristotelismus bei den Griechen*, vol. 2 (Berlin)

Rashed, M. (1997) 'A "new" text of Alexander on the soul's motion', in R.R.K. Sorabji (ed.) *Aristotle and After*, BICS suppl. vol. 68.

Riet, S. van (1991) 'A propos de la biographie de Simplicius', *Rev. Phil. de Louv.* 89, 506-14

Robinson, R. (1953[2]) *Plato's Earlier Dialectic* (Oxford)

——— (1969) *Essays in Greek Philosophy* (Oxford)

Ross, W.D. (1936) *Aristotle: Physics* (Oxford)

——— (1949) *Aristotle's Prior and Posterior Analytics* (Oxford)

Solmsen, F. (1960) *Aristotle's System of the Physical World: A Comparison with his Predecessors* (Ithaca NY)

Sorabji, R.R.K. (1982) *Time, Creation and the Continuum* (London)

——— (1988) *Matter, Space and Motion* (London)

Sorabji, R.R.K. (ed.) (1987) *Philoponus and the Rejection of Aristotelian Science* (London)

——— (ed.) (1990) *Aristotle Transformed* (London)

Tardieu, M. (1986) 'Témoins orientaux du *Premier Alcibiade* à Harrân et à Nag 'Hammâdi', *Journal Asiatique* 127

——— (1986) 'Sâbiens coraniques et "Sâbiens" de Harrân', *Journal Asiatique* 274, 1-44

——— (1987) 'Les calendriers en usage à Harrân d'après les sources arabes et le commentaire de Simplicius à la *Physique* d'Aristote', in Hadot, 1987

Tricot, J. (1949) *Aristoteles: Traité du Ciel* (Paris)

Verrycken, K. 'The development of Philoponus' thought and its chronology', in Sorabji, 1990

Wehrli, F. (1969) *Die Schule des Aristoteles* (Basel)

Wildberg, C. (1987) *Philoponus: Against Aristotle on the Eternity of the World* (London)

——— (1988) *John Philoponus' Criticism of Aristotle's Theory of Ether* (Berlin and New York)

Williams, C.J.F. (1982) *Aristotle: On Generation and Corruption*, Clarendon Aristotle Series (Oxford)

Wolff, M. (1987) 'Philoponus and the rise of pre-classical dynamics', in Sorabji, 1987

English-Greek Glossary

account: *logos*
activity: *energeia*
actuality: *energeia*
adduce: *epagein*
aether [fire]: *aithêr*
age: *aiôn*
ageless: *agêrôn*
aim: *skopos*
air: *aêr*
alteration: *alloiôsis*
animal: *zôion*
area: *platos*
argue: *epagein*
argument: *epikheirêma*, *epikheirêsis*
argument: *logos*
art: *tekhnê*
assumption: *hupokeimenon*
atom: *atomos*
axiom: *arkhê*, *axiôma*

beginning of motion: *arkhê kinêseôs*
beginning: *arkhê*
body: *sôma*

carried round, be: *peripheresthai*
carried, be: *pheresthai*
cause: *aition*
centre: *kentron*
change: *kinêsis*
change: *metaballein*
change: *metabolê*
circle: *kuklos*
circular motion: *kuklôi kinêsis*
circumference: *periphereia*
cold (n.): *psuxis*
come to be: *gignesthai*
commensurable: *summetros*
common: *koinos*
commonality: *koinotês*
complete actuality: *entelekheia*
complete (adj.): *teleios*
composite: *sunamphoteron*, *sunthetos*
composition: *sustasis*
compound: *sunthetos*
conclude: *epagein*, *sumperainein*, *sunagein*
conclusion: *sumperasma*
concrete: *pragmatikos*
conditional: *sunêmmenon*
construct: *sunistanai*
contain: *periekhein*
continuous: *sunekhes*
contrariety: *enantiôsis*
contrary: *enantion*
conviction: *pistis*
cut: *temnein*

demonstrate: *apodeiknunai*
demonstration: *apodeixis*
dense: *puknos*
describe a circle: *kuklographein*
destructible: *phthartos*
destruction: *phthora*
determinate: *hôrismenos*
determine: *hôrizein*
determined: *aphôrismenos*
difference: *diaphora*
difficulty: *aporia*
diminution: *meiôsis*
disagreement: *enantiôsis*
disposition: *diathesis*
distance: *apostasis*, *diastasis*, *diastêma*
divine: *theios*
division: *diairesis*, *diastasis*
draw a consequence: *epagein*
earth: *gê*
element: *stoikheion*
enclose: *periekhein*
end: *telos*
ensemble: *holotês*
essence: *to einai*
eternal: *aidios*
evidence: *marturia*

exceed: *huperballein*
exemplar: *paradeigma*
explain: *didaskein*
extension: *paratasis*
external mover: *exôthen kinoun*
extremity: *eskhaton*
extrusion: *ekthlipsis*

fifth body: *pempton sôma*
final cause: *telikon aition*
finite body: *peperasmenon sôma*
finite distance: *diastasis peperasmenê, diastêma peperasmenon*
finite time: *khronos peperasmenos*
finite: *peperasmenos*
fire: *pur*
fire-sphere: *hupekkauma*
first cause: *prôton aition*
forcible: *biaios*
form: *eidos*

generable: *genêtos*
generally: *katholou*
generated: *genêtos*
generation: *genesis*
genus: *genos*
get clear of: *apoluesthai*
getting clear: *apolusis*
God: *theos*
gold: *khrusos*
growth: *auxêsis*

heat: *thermotês*
heaven: *ouranos*
heavenly bodies: *ourania*
heavenly: *ouranios*
heaviness: *barutês*
heavy: *barus*

immobile cause: *akinêton aition*
immortal: *athanatos*
immortality: *athanasia*
impossible: *adunatos*
inclination: *rhopê*
incommensurable: *asummetros*
incorporeal: *asômatos*
indestructible: *aphthartos*
indeterminacy: *aoristia*
indeterminate: *aoristos*
indication: *tekmêrion*
individual: *atomos*
indivisible: *adiairetos, atomos*
induction: *epagôgê*
infer: *epagein*
infinite: *apeiros*
intelligible: *noêtos*
intermediate: *metaxu*
interval: *diastêma*
investigate: *zêtein*

light (adj.): *kouphos*
lightness: *kouphotês*
line: *grammê*

magnitude: *megethos*
matter: *hulê*
measure out: *katametrein*
measure: *metrein, metron*
middle: *meson*
minor premiss: *proslêpsis*
motion: *kinêsis*
motionless: *akinêtos*
move in circle: *kuklôi kineisthai*
move in opposite directions: *antikineisthai*

natural body: *phusikon sôma*
natural motion: *kinêsis phusikê*
natural philosopher: *phusiologos*
natural place: *topos kata phusin*
natural scientist: *phusikos*
naturally constituted: *phusei sunestôta*
non-uniform: *anomoiomerês*
number: *arithmos, plêthos*

objection: *enstasis*
one, the: *to hen*
orbit: *periphora*
outermost heaven: *eskhatos ouranos*

part: *meros, morion*
participate: *metekhein*
participation: *methexis*
particle: *morion*
partless: *amerês*
perceptible: *aisthêtos*
place: *topos*
plane: *epipedon*
power: *dunamis*
premiss: *protasis*
primarily: *proêgoumenôs*
prime mover: *prôton kinoun*
principle: *arkhê*
productive cause: *poiêtikon aition*

proof: *deixis*
proper place: *oikeios topos, topos oikeios*
proportion: *analogia, logos*

quick: *takhus*

rare(fied): *manos*
reason: *aitia*
reciprocal motion: *antiperistasis*
remain at rest: *menein*
rest: *monê, stasis*
revolution: *kuklophoria, peristrophê*
rise: *epipolazein*

self-moving: *autokinêtos*
separable: *khôrizomenos*
separation: *diastasis*
settle: *huphistasthai*
shape: *skhêma*
simple body: *haploun sôma*
simple motion: *haplê kinêsis*
soul: *psukhê*
source of motion: *arkhê kinêseôs*
source: *arkhê*
species: *eidos*
speed: *takhutês*
sphere of the fixed [stars]: *aplanês sphaira*
sphere: *sphaira*
squeezed out, be: *ekthlibesthai*
starting-point: *arkhê*
subject: *hupokeimenon*
sublunary: *to hupo selênên*
substance: *ousia*
substitute: *metalambanein*
substrate: *hupokeimenon*
subtract: *aphairein*
subtraction: *aphairesis*
supposing: *hupothemenos*
supposition: *hupothesis*
syllogism: *sullogismos*
symmetry: *summetria*

time: *khronos*
topic: *topos*
treatise: *pragmateia*

unaffectible: *apathês*
unceasing: *apaustos*
unchangeable: *ametablêtos*
ungenerated: *agenêtos*
unified: *hênômenos*
uniform in nature: *homophuês*
uniform: *homoiomerês*
unnatural motion: *kinêsis para phusin*
unnatural place: *topos para phusin*
untraversable: *adiexitêtos*

void: *kenon*

water: *hudôr*
weight: *to baros*
world: *kosmos*

Greek-English Index

References are to the page and line numbers of Heiberg's *CAG* edition of 1894, which appear in the margin of the translation.

Subject Index